Ertl/Küppers
Low Energy Electrons
and Surface Chemistry

Monographs in Modern Chemistry

Series Editor: Hans F. Ebel

Vol.1: F. Kohler
The Liquid State

Vol.2: H. Meier
Organic Semiconductors

Vol.3: H. Kelker/R. Hatz
Liquid Crystals

Vol.4: G. Ertl/J. Küppers
Low Energy Electrons
and Surface Chemistry

Vol.5: E. Breitmaier/W. Voelter
^{13}C NMR Spectroscopy

Vol.6: D.O. Hummel (Editor)
Polymer Spectroscopy

Vol.7: D. Ginsburg
Propellanes

The series is to be continued

ERTL/ KÜPPERS

◇ LOW ENERGY ELEC- TRONS AND SURFACE CHEMI- STRY

VERLAG CHEMIE

Prof. Dr. Gerhard Ertl
Physikalisch-Chemisches
Institut der Universität
8 München 2
Sophienstrasse 11

Doz. Dr. Jürgen Küppers
Physikalisch-Chemisches
Institut der Universität
8 München 2
Sophienstrasse 11

Copy Editing: Dr. Gerd Giesler

This book contains 152 figures and 3 tables

ISBN 3-527-25562-1

Library of Congress Catalog Card No. 74-82758

© Verlag Chemie, GmbH, D-694 Weinheim, 1974

Composition: Helmut Becker-Filmsatz, D-6232 Bad Soden;
Printed by Alexander Krebs, D-6944 Hemsbach; Bookbinder: Hollmann KG., D-6100 Darmstadt

Printed in Germany

Preface

The study of processes at solid surfaces is at present a rapidly growing field of chemistry and physics which was mainly initiated by the development of a series of experimental techniques based on the interaction properties of low energy electrons with solid matter.

At the time we started to write this monograph a very large number of original publications and also a series of review papers covering particular branches were available, but we felt that there was some need for a more comprehensive treatment with an introductory character for non-experts in the field. No attempt was therefore made to present a complete review or to give a full outline of the theoretical aspects of the techniques. The intention was rather to give a description of the possible applications of the methods to problems of surface chemistry (mainly adsorption) than to present a thorough discussion of the physical background. Furthermore this volume does not contain a series of important techniques which are based not on the interaction of slow electrons with matter but on other types of radiation. The main emphasis is put on the information which is accessible on the chemical, geometric and electronic properties of the topmost atomic layers of solids. The most important techniques in this respect are Auger electron spectroscopy, low energy electron diffraction and photo-electron spectroscopy respectively. Since each method illuminates different aspects of the surface phenomena the present trend is to combine as many of these techniques as possible in the same apparatus, in particular since a complete theoretical analysis of the data is – due to the complicated primary processes involved – not yet possible.

Surface research today is mainly characterized by the 'clean single crystal surface' approach. Although this situation is far from that for 'real' surfaces it is felt that this is the only way to learn something about the elementary steps of surface processes involved in such complex phenomena as catalysis.

The manuscript for this volume was completed in the summer of 1973 and it therefore covers the literature until about the end of 1972 although a few of the more recent results were added in proof. We are heavily indebted to Dr. C. Snaith who considerably improved the English version and to Dr. O. Schober for a critical reading and for compiling the index.

München, April 1974
G. Ertl
J. Küppers

Contents

1. Basic concepts 1
1.1. Introduction 1
1.2. Principles of ultrahigh vacuum technique 2
1.2.1. Why is UHV necessary? 2
1.2.2. Production of ultrahigh vacuum 3
1.2.3. Pressure measurements 3
1.2.4. Gas handling 4
1.3. Preparation of clean surfaces 4
1.4. Interactions of slow electrons with matter 7
1.5. Electron energy analyzers 9
1.5.1. Retarding field grid analyzer (RFA) 9
1.5.2. Cylindrical mirror analyzer (CMA) 12
1.5.3. 127°-analyzer 13
1.5.4. Concentric hemisphere analyzer (CHA) 14
1.6. References 15

2. Auger electron spectroscopy 17
2.1. Historical development 17
2.2. Experimental 18
2.2.1. The source of excitation 18
2.2.2. The sample 19
2.2.3. Analyzer and detector system 19
2.2.4. Further refinements 22
2.3. Mechanism of the Auger process 24
2.4. Energies and shapes of the Auger peaks 30
2.4.1. Free atoms 30
2.4.2. Condensed matter 30
2.4.3. Chemical effects 33
2.5. Intensity of the Auger electron emission 34
2.5.1. The Auger yield 34
2.5.2. The ionization cross section 36
2.5.3. Auger electron emission from condensed matter 37
2.6. The detected volume 39
2.7. Qualitative analysis 41
2.8. Quantitative analysis 41
2.8.1. Determination of relative surface quantities 41
2.8.2. Absolute surface quantities 42
2.8.3. Alloys 43
2.8.4. Depth profiling 45
2.8.5. Kinetic studies 46
2.9. Deconvolution technique and band structure 47
2.10. References 49

3. Electron Energy Loss Spectroscopy 53
3.1. General remarks 53
3.2. Experimental 54
3.3. Ionization spectroscopy 54
3.4. Plasmon losses 57
3.5. Inelastic low energy electron diffraction (ILEED) 60
3.6. Changes of energy-loss spectra by adsorbed layers 62
3.7. References 65

4. Photoelectron Spectroscopy 67
4.1. Introduction 67
4.2. Experimental 69
4.2.1. Light sources 69
4.2.2. Analyzer and detector 70
4.3. Photoelectron emission from solids 72
4.4. Band structure of metals and alloys 77
4.5. Metal surfaces and chemisorption levels 79
4.6. Semiconductors 82
4.7. References 83

5. Appearance Potential Spectroscopy (APS) 85
5.1. Introduction 85
5.2. Experimental 86
5.3. The mechanism of APS 88
5.4. Surface analysis 91
5.5. Energies of atomic core levels 91
5.6. Band structure of metals and alloys 93
5.7. Plasmon coupling 95
5.8. Chemical effects 96
5.9. References 97

6. Field Emission Spectroscopy 99
6.1. Introduction 99
6.2. Energy distribution of field emitted electrons from clean surfaces 100
6.3. Adsorbed layers and resonance tunneling 103
6.4. References 106

7. Ion Neutralization Spectroscopy 109
7.1. References 114

8. Work function and contact potential 115
8.1. Introduction 115
8.2. Experimental techniques 117
8.2.1. Thermionic electron emission 117
8.2.2. Photoemission 118
8.2.3. Field emission 118
8.2.4. Field emission retarding potential method (FERP) 119
8.2.5. Vibrating capacitor method (Kelvin method) 120
8.2.6. Diode method 121

8.3. The work function of clean surfaces 122
8.4. Adsorbed layers 123
8.5. References 126

9. Low energy electron diffraction (LEED) 129
9.1. Historical development 129
9.2. Classification of surface structures 130
9.3. Formation of the diffraction pattern 133
9.4. Experimental equipment 135
9.4.1. Introduction 135
9.4.2. Electron gun 135
9.4.3. Crystal manipulator 136
9.4.4. Detector system 137
9.5. Geometrical theory of diffraction 140
9.5.1. Introduction 140
9.5.2. The reciprocal lattice 141
9.5.3. Interference conditions and the Ewald construction 143
9.5.4. Analysis of a simple diffraction pattern 145
9.5.5. Domain structures 146
9.5.6. LEED pattern of coincidence lattices and incoherent structures . . . 149
9.6. Kinematic theory 152
9.6.1. Introduction 152
9.6.2. Scattering at two-dimensional lattices 152
9.6.3. Kinematical structure factor 156
9.6.4. Intensity-voltage (I/U) curves 157
9.7. Disordered structures 159
9.7.1. Introduction 159
9.7.2. The diameter of the coherence zone 159
9.7.3. Size effects 160
9.7.4. Substitution disorder 161
9.8. Facets and stepped surfaces 166
9.8.1. Faceting 166
9.8.2. Surfaces with regular steps arrays 168
9.9. Simulation of diffraction patterns 170
9.10. Elements of a dynamic LEED theory 171
9.10.1. The need for a dynamic theory 171
9.10.2. Physical parameters entering a complete theory 173
9.10.3. Semi-empirical extensions of the kinematic theory 175
9.10.4. Multiple-scattering models 176
9.11. Averaging methods 179
9.12. Structural aspects of clean surfaces 182
9.12.1. Atomic distances in surface layers 182
9.12.2. Metals 183
9.12.3. Alloys 185
9.12.4. Semiconductors and semimetals 185
9.12.5. Ionic crystals and insulators 187
9.13. References 188

10. Vibrations at Surfaces. 193
10.1. Introduction. 193
10.2. The vibrating lattice/temperature effects in LEED 194
10.2.1. Variations of the LEED-intensities 194
10.2.2. Shifts of the Bragg maxima. 201
10.2.3. Thermal diffuse scattering 202
10.3. High resolution energy loss spectroscopy. 205
10.4. Infrared spectroscopy 207
10.5. References 208

11. Processes in adsorbed layers 211
11.1. Introduction. 211
11.2. Reactions of electrons with adsorbed particles 213
11.3. Ordered adsorbed phases 216
11.4. Physical adsorption. 220
11.5. Metallic adsorbates 221
11.6. Chemisorption of gases 225
11.7. Co-adsorption 230
11.8. Reactions at surfaces 232
11.9. References 236

Index 239

1. Basic concepts

1.1. Introduction

Processes occurring at solid surfaces are of great practical importance particularly for the study of heterogeneous catalysis, corrosion, semiconductor technology etc. Basic research in solid state science is increasingly confronted with problems connected with surface effects since all solids have boundaries. Our knowledge of surface properties is much inferior to that of ordinary threedimensional phases due to the additional experimental and theoretical complications associated with the absence of the third dimension. In particular we are confronted with a problem of quantity: one square centimeter of a surface contains only about 10^{15} particles or 10^{-9} mole. In practice this difficulty is usually countered by working with materials which are porous or consist of very small particles giving a high surface to volume ratio. Unfortunately such systems are often unsuitable for investigations into primary processes since the surfaces consist of many planes of different crystallographic orientation with structural defects and of indeterminate chemical composition.

One can visualize a surface as the result of a fracture along some plane in the bulk material, the bonds between neighbouring atoms being simply parted. It is then reasonable to imagine the surface as consisting of more or less unsaturated valencies which show a strong tendency to interact chemically with particles arriving from the gas phase.

In order to maintain the original state for long enough to conduct experiments it is clearly necessary to keep the pressure of the residual gas above the surface very low, which has only become possible since the development of ultra high vacuum techniques. Conversely the lowest attainable pressures are limited by desorption of gases from the surfaces within the vacuum system, so that surface and vacuum research are intimately related to each other.

Two limiting cases are usually considered when describing the properties of three dimensional matter: the ideal gas and the ideal solid. The former is composed of point particles which interact only by elastic collisions. The latter consists of a strictly periodic arrangement of atoms forming a lattice without defects or impurities, extending to infinity in all three dimensions. Neither can actually be realized, but these models are nevertheless useful since their properties can be described relatively simply. The ideal laws so derived are frequently applied to real systems by either bearing in mind their limitations or by modifying them on a semi-empirical basis. This approach can also be used when considering the properties of solid surfaces. An 'ideal solid surface' is defined by analogy with the 'ideal solid' and consists of a perfectly periodic arrangement of surface atoms in two dimensions (extended to infinity in these directions) without any structural defects or foreign atoms. As will be shown it is frequently advantageous to consider not only the topmost atomic layer but also to include a limited number of deeper layers in a complete treatment of the surface region. The three dimensional periodicity and the bulk properties are generally not cut off abruptly at the surface, but a superficial region (perhaps one or a few atomic layers thick) is to be expected where the geometric, chemical, and electronic properties may be different from those of the bulk. As proposed by May [1] we define the substrate surface as the plane below which the threedimensional periodicity and all the other properties of the substrate apply, and the 'surface' as that region above this plane where these properties are different. The 'substrate surface' thus acts as an interface between the surface region and the bulk.

The chemist is in principle interested mainly in substantial changes in the systems under investigation, whereas the physicist is concerned more with the properties of the system itself. In surface science this frontier between physics and chemistry is almost nonexistent. There are however some problems which are mainly physical in nature and which therefore are not treated extensively here. The investigation of surface processes by studying the geometric, chemical, and electronic structure of clean surfaces including their dynamic properties will be considered together with interactions with the gas phase (e.g. adsorption, epitaxy, and heterogeneous catalysis).

Several experimental methods for the investigation of such problems have been developed during the past years using low energy electrons ($E \leqslant 1\,000$ eV). Electrons in this energy range interact with matter very strongly, and their mean free path within solids is consequently only a few atomic layers, making this an ideal tool for surface research. The main purpose of this monograph is to describe these methods, their limitations and their application to problems of surface chemistry.

1.2. Principles of ultra high vacuum technique [2]

1.2.1. Why is UHV necessary?

As already mentioned a clean surface is normally very reactive towards particles impinging upon it from the gas phase. In order to maintain clearly defined surface conditions it is therefore necessary to reduce the rate at which gas molecules approach until the time for the surface to become completely covered by adsorbed gas is much longer than that required for the experiment. It can be shown directly from the kinetic gas theory that the number $\dot{n}_s$ of particles striking a surface of 1 cm^2 in 1 s is given by

$$\dot{n}_s = \tfrac{1}{4} N_g \bar{v} \tag{1.1}$$

where N_g is the number of gas molecules per cm^3 and $\bar{v}$ their mean thermal velocity. From this follows that

$$\dot{n}_s = N_g \sqrt{\frac{RT}{2\pi M}} \approx 3.5 \cdot 10^{22} \, \frac{p}{\sqrt{MT}} \; [\mathrm{cm}^{-2}\,\mathrm{s}^{-1}] \tag{1.2}$$

with $R =$ gas constant, $T =$ absolute temperature, $M =$ molecular weight, and $p =$ gas pressure in Torr.

Assuming a monolayer capacity of $3 \cdot 10^{14}$ particles/cm^2, an average molecular weight of $M = 28$, and $T = 300$ K, eq. (1.2) yields

$$\dot{n}_s \approx 10^6 \cdot p \; [\mathrm{monolayers/s}] \tag{1.3}$$

which means that at a pressure of 10^{-6} Torr (standard high vacuum conditions) the number of molecules necessary for the build-up of a monolayer is offered to a surface every second. The time of coverage τ_c then depends on the sticking coefficient s, i.e. the probability that an impinging molecule becomes adsorbed:

$$\tau = \frac{1}{\dot{n} \cdot s} \approx \frac{10^{-6}}{s \cdot p} \tag{1.4}$$

s is quite frequently close to unity. In order to obtain a value of 1 hour for τ_c (the time necessary for performing an experiment with a 'clean' surface) residual gas pressures in the range of 10^{-10} Torr must be reached. Ultrahigh vacuum (UHV) is thus essential for surface work.

1.2.2. Production of ultrahigh vacuum

The best vacuum attainable in UHV systems is determined mainly by gas emitted from the materials used to build the apparatus. The initial evolution of gas from adsorbed layers can be accelerated considerably by heating, so that extended bake-out periods at temperatures up to 400 °C are usually necessary. Most of the vacuum systems currently in use are of stainless steel components joined together by flanges with metal seals (copper or gold). Systems of borosilicate glass (e.g. Pyrex, Duran) are also used, and frequently larger systems are partly constructed of glass with a bakeable metal to glass seal connection to the stainless steel section. In most cases pressures of the order of 10^{-10} Torr are attained by ion getter pumps with pumping speeds from 50 to 200 l/s. The ability (or inability) of such pumps to pump noble gases is important, as is their tendency to emit gases by themselves, i.e. the pumping speed varies considerably with the kind of gas and may also be influenced by the history of the system. The great advantage of these pumps however is that they need practically no maintenance and that they allow UHV to be simply and routinely obtained. The best results are achieved by combining an ion getter pump with a titanium sublimation pump, in which freshly evaporated Ti surfaces develop phenomenal pumping speeds in the lower pressure ranges.

Turbo-molecular pumps may find increasing use in the future since they are able to handle large quantities of gas, noble gases included, which is necessary if the system contains a gas discharge lamp and no separating window for example as is the case with ultraviolet photoelectron spectrometers.

Mercury and oil diffusion pumps are in use in some laboratories and allow pressures in the 10^{-10} Torr region to be obtained without undue effort. Their major disadvantage is the danger of surface contamination by pump fluid vapours. Particular attention must therefore be paid to the maintenance of suitable traps and baffles.

The use of cryopumps is comparatively rare although when operated with liquid helium they can provide the cleanest ultrahigh vacuum. The running costs of such systems are still rather high.

UHV pumps cannot pump at atmospheric pressure so that a forevacuum or roughing system is invariably required. Sorption pumps or rotary oil pumps are commonly used for this purpose.

1.2.3. Pressure measurement

Pressure is usually measured as the total pressure within the system. When working with ion getter pumps this can be done simply by measuring the current flowing through the pump, but this crude method gives only an estimate. Most total pressure measurements in the range between 10^{-3} and 10^{-10} Torr are made with Bayard-Alpert ionisation gauges. These are relatively inexpensive and easy to operate. Their major disadvantage is that they can influence the system under investigation; for example in the ionisation process some molecules are withdrawn from the gas phase so that a small pumping effect results. The presence of the hot filament is even more troublesome in that the temperatures required for emission are capable of dissociating gases. This effect is particularly marked for hydrogen, and the atoms so produced react readily with parts of the apparatus or with other components within the gas phase. Molecules may also decompose and react with impurities in the filament, for example the CO partial pressure in the apparatus may increase considerably after switching on the ion gauge owing to the presence of carbon on the tungsten filament.

It is consequently of fundamental importance to monitor the composition of the gas phase with a small mass spectrometer [3], even in those cases where a partial pressure analysis is not of interest *a priori*.

Quadrupole mass filters are almost ideal for this purpose. They are quite small and need no permanent magnet and can therefore be arranged in any position in order to bring the ionisation source as near as possible to the surface under investigation, which is particularly useful for studies of desorption or catalytic processes with a surface of less than 1 cm^2 in area. These instruments have high sensitivity and are typically able to detect partial pressures of as low as 10^{-14} Torr.

The Omegatron mass spectrometer is still in use in some laboratories. The spectrometer head is small but requires a large permanent magnet. In one version [4] it is possible to heat the metal parts of the gauge during measurement which reduces the sorption capacity and the 'memory effect' to minimum values.

1.2.4. Gas handling

The gases used in clean surface studies are usually contained in glass flasks with break-off seals and are introduced into the vacuum system by means of bakeable dosing valves which allow controlled inlet rates into UHV direct from atmospheric pressure. Measurements are then conducted in a static flow system, where a constant pressure is maintained by opening the leak valve and throttling the pumping speed. The other possibility is to expose the surface to a known gas pressure for a measured period of time, which is useful in cases of irreversible adsorption. 1 Langmuir [L] = 10^{-6} Torr · s has been proposed as unit for this type of exposure [5]. If the sticking coefficient is unity, 1 L corresponds to the number of impinging gas molecules which is just necessary to build-up around a monolayer (if no desorption occurs).

The commercially available gases are usually of a very high degree of purity so that further purification is unnecessary. H_2 and O_2 can be cleaned and admitted into vacuum by diffusion through heated Pd and Ag respectively. Methods for purifying gases have been reviewed by Mobley [6]. It is frequently possible to reduce the impurity content to 1 part per million, but once admitted into the vacuum chamber the purity of the gas is subject to change by interaction with gases adsorbed on parts of the UHV system or with hot filaments.

1.3. Preparation of clean surfaces

The conditions for the definition of a clean surface are at present much less strict than for three dimensional phases. A surface is called clean if its impurity content is below the detection limit of current chemical analysis methods. A typical value is about 1 % of a monolayer, but it is still somewhat uncertain whether or not still smaller amounts of surface contaminants can give rise to noticeable effects. (This will certainly be the case for the electrical properties of semiconductor surfaces.) The cleanliness of the surface has only been the subject of direct measurement since the development of Auger electron spectroscopy (1968), at which time it became clear that even more care was necessary in the preparation and maintenance of clean surfaces than was previously thought. No generally applicable recipe for the preparation of clean surfaces can be given. In each case the proper treatment must be determined experimentally. This is frequently more time-consuming than the subsequent investigations on the clean surface.

The following techniques have been shown to be useful for preparing clean surfaces:

a) Cleaving in ultrahigh vacuum. Brittle materials, especially semiconductors, may be cleaved along certain crystallographic orientations which produces smooth surfaces with areas of a few mm^2 [7]. An arrangement for cleaving in UHV which also allows cooling of the sample has been described by Palmberg [8]. It can be assumed that this technique produces surfaces whose degree of cleanliness is initially as good as that of the bulk. However this method is restricted to certain surface orientations of a few materials. If there is no need to produce a single crystal surface with distinct orientation, cleaved surfaces can be obtained much more simply by crushing thin slabs of the material with a magnetically operated hammer in UHV [9]. Large total areas can be obtained in this way which can be used for catalytic studies for example [10].

b) High temperature treatment. Previously it was assumed that clean surfaces of the refractory metals e. g. W or Mo could be prepared by heating to very high temperatures, thereby desorbing all impurities. However it became evident that at high temperatures impurities which are dissolved in the bulk can diffuse to the surface, so that annealing at high temperatures may lead to an accumulation of contaminants instead of removing them. Sulfur and carbon are the major culprits here. In short, the production of clean surfaces by heat treatment alone is difficult, if not impossible.

c) Chemical reactions. A combination of surface reactions can be useful in some cases. For example it has been demonstrated [11] that carbon and sulfur impurities on Pd surfaces can be oxidized away directly with oxygen at higher temperatures and that the adsorbed oxygen can be removed easily with CO which forms CO_2 and desorbs. The residual CO desorbs above 200 °C.

To remove carbon from Ni surfaces a combined oxidation/reduction treatment using H_2 and O_2 has been applied successfully [12]. An interesting combined chemical reaction/sputtering method has been used recently by Tracy [13] in order to obtain a clean Ni surface: Carbon reacts with oxygen to form volatile CO and CO_2, and the absorbed oxygen may be removed by interaction with K^+ ions originating from an alumino-silicate ion source.

In contrast to earlier opinions clean tungsten surfaces cannot be produced by heating alone [14]. However the carbon which accumulates at high temperature at the surface can be removed by reaction with oxygen, and subsequent heating then produces clean surfaces provided that during this process further contaminants are not allowed to diffuse to the surface [15].

d) Field desorption [16]. This method is normally restricted to very sharp tips with diameters of about 1 000 Å as used for field emission studies. The action of very high electric fields can cause the evaporation of surface atoms, but again combination with a chemical treatment may be necessary.

e) Evaporated thin films [17]. Evaporation and condensation of metals is in principle a simple method for the preparation of clean surfaces with a large area. However this technique is also prone to diffusion of impurities from the bulk of the evaporator and care must be taken to avoid distillation of these onto the fresh surface. The production of single crystal surfaces is somewhat more difficult and is most conveniently achieved by epitaxial growth on substrates like mica or rocksalt. The most favourable substrate temperatures and evaporation rates must be determined experimentally [18]. It is even possible to prepare clean single crystal surfaces of alloys with varying compositions by this technique [19].

f) Noble gas ion sputtering. Sputtering of the topmost atomic layers of crystals by bombardment with noble gas ions (usually argon) was first applied by Farnsworth et al. in 1958 [20] to prepare clean surfaces. This method is without doubt the most versatile and has nearly universal applicability. Noble gas ions are produced by electron impact with gas atoms which are admitted to the vacuum chamber at a pressure of $\leqslant 10^{-3}$ Torr. The ions are then accelerated by a voltage of a few hundred volts towards the sample surface. The ion current and duration of the bombardment depend on the kind of material and on the thickness of the layer to be removed. Fig. 1.1 shows a commercially available ion gun of this type. The sample is positioned

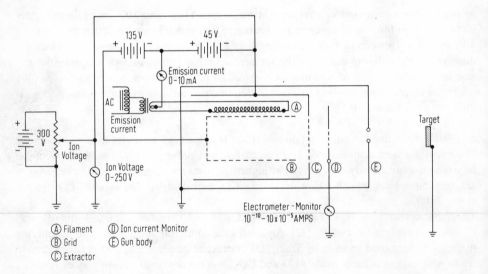

ⒶFilament   ⒹIon current Monitor
ⒷGrid   ⒺGun body
ⒸExtractor

Fig. 1.1. Noble gas ion gun (Varian).

in front of the opening of the gun. Some guns are fitted with focussing and deflection plates for accurate positioning of the ion beam, e.g. for sputtering particular areas of the sample. Ion bombardment guns are to be found on most modern surface research systems.

The lattice of the crystal suffers damage in the surface region from the ion bombardment and furthermore noble gas atoms may be incorporated. The ion bombardment must therefore be followed by a recrystallization process of careful annealing. It is possible that during this heating treatment impurities from the bulk again diffuse to the surface, so that the ion bombardment/annealing cycle has to be repeated several times. Sometimes it is favourable to keep the sample at elevated temperatures during the ion bombardment. It is of particular importance to minimize the partial pressures of residual gases (mainly CO) during the sputtering process since these molecules may otherwise become adsorbed or implanted into the lattice after ionization in the ion gun. This can be achieved by operating a small getter pump during sputtering since such pumps have a negligible speed for noble gases.

1.4. Interactions of slow electrons with matter

Electrons with energies between about 10 and 1000 eV are ideally suited to investigate the topmost layers of solids because their mean free path in solids is of the order of only a few atomic layers. The characteristic dependence of this property on the electron energy can be seen in fig 1.2. Although it is somewhat difficult to obtain exact quantitative data the general

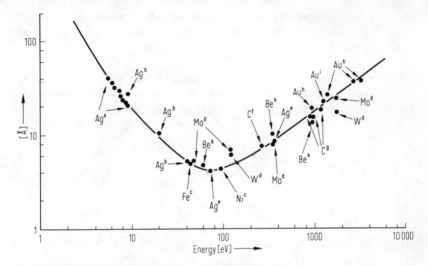

Fig. 1.2. Mean free path of electrons in metallic solids as a function of their energy.
a: H. Kanter, Phys. Rev. *B1*, 522 (1970) (Electron transmission). b: D.E. Eastman, 32nd Physical Electronics Conference, Albuquerque, N. Mex. 1972 (UPS). c: J. W. T. Ridgeway and D. Haneman, Surface Sci. *24*, 451 (1971); *26*, 683 (1971) (AES). d: M. L. Tarng and G. K. Wehner, J. appl. Phys. *43*, 2268 (1972) (AES). e: P. W. Palmberg and T. N. Rhodin, J. appl. Phys. *39*, 2425 (1968) (AES). f: K. Jacobi and J. Hölzl, Surface Sci. *39*, 54 (1971). g: R. G. Steinhardt, J. Hudis and M. L. Perlman, in: Electron Spectroscopy (D.A. Shirley, ed.) North Holland, Amsterdam (1972), p. 557 (XPS). h: M. Klasson, J. Hedman, A. Berndtson, R. Nilsson and C. Nordling, Physica Scripta *5*, 93 (1972) (XPS). i: Y. Baer, P.F. Heden, J. Hedman, M. Klasson and C. Nordling, Solid State Comm. *8*, 1479 (1970) (XPS). k: M. P. Seah, Surface Sci. *32*, 703 (1972) (AES).

features have been well established as shown in the figure. The mean free paths are also a function of the substrate concerned to some extent. A minimum in the mean free path of only a few Ångstroms occurs at energies between 40 and 100 eV. It increases steeply at energies below 20 eV, but in the direction of higher energies the mean free path increases only slowly and at 1 000 eV is still only a few atomic layers. Further data were recently reviewed by Brundle [34].

This result forms the basis of a whole series of methods for the investigation of surface properties. These techniques may be divided in the following three groups:

a) Surface excitation: low energy electrons, measured response: non electronic. A surface bombarded with slow electrons emits electromagnetic radiation which originates from the surface layers even though the electromagnetic radiation itself has a much larger free path in solid matter. Soft x-ray appearance potential spectroscopy (APS) is one example of this type of technique. Another possibility consists of examining particles (atoms, molecules or ions) desorbing from an electron bombarded surface which is covered by an adsorbate.

b) Surface excitation: non electronic, measured response: the emission of low energy electrons. The most important technique of this type is photoelectron spectroscopy, where electromagnetic radiation is used for the excitation of bound electrons. The small escape depth of the electrons makes this a surface-sensitive method, although the primary radiation has a larger depth of penetration. Other techniques belonging in this category are field electron emission and ion neutralisation spectroscopy.

c) Excitation and response both low energy electrons. A beam of electrons with a definite energy E_p impinging on a surface will give rise to the appearance of backscattered and secondary emitted electrons whose energy distribution is schematically shown in fig. 1.3. This distribution

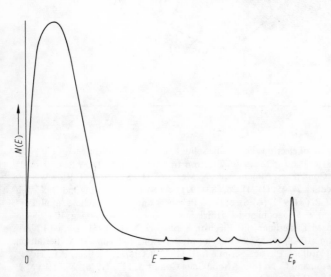

Fig. 1.3. Energy distribution $N(E)$ of back-scattered slow electrons as a function of their energy. E_p is the energy of the primary electrons.

may be divided into the following three regions:

i) The so-called "true secondary electrons" are located in the very low energy range and are created as a result of inelastic collisions between primary electrons and electrons bound in the solid. In each collision process only a relatively small amount of energy is transferred, so that a single primary electron can create several secondary electrons which then appear in the lower part of the energy scale. Although the yield and energy distribution of secondary electrons is strongly influenced by adsorbed layers a systematic attempt to analyse the true secondary electrons in view of the problems of surface chemistry has yet to be made.

ii) The medium energy range is characterized by a relatively smooth background on which small peaks are superposed. These are caused primarily by the emission of Auger electrons (→ Auger electron spectroscopy) and by characteristic energy losses of the primary electrons due to single and collective excitation of the valence and core electrons in the solid (→ electron energy loss spectroscopy).

iii) Some small fraction of the primary electrons (typically a few percent) is back-scattered from the surface elastically without any noticeable energy loss. In practice the width of this elastic peak is determined by the energy spread of the primary electrons and by the limited resolution of the energy analyzer. Normally the so-called 'quasi-elastic' electrons which have suffered energy losses of much less than 1 eV by interaction with phonons in the surface region are not resolved, but by using an energy analyzer of very high resolution and a highly monochromatic primary electron beam it is possible to separate the elastic and quasi-elastic electrons, which may be used to analyze surface vibrations. Electrons elastically scattered from a single crystal can interfere at periodic atomic arrangements on the surface because of their de Broglie-wave properties. This is referred to as low energy electron diffraction (LEED). The method has the same relevance in surface crystallography as x-ray diffraction techniques have for the structural analysis of three-dimensional matter.

1.5. Electron energy analyzers

As is evident from the preceding outline of experimental methods most of them are based on an analysis of the energy distribution of electrons emitted from a surface. The different features of the emission spectra (type of emission, energy range, resolution needed, intensity available etc.) have lead to the development of several different analysis techniques which are described separately in the following chapters. But most of them have in common that they need an analyzer for measuring the energy distribution of the emitted electrons. These analyzers are (at least in principle) compatible with all the different methods.

1.5.1. Retarding field grid analyzer (RFA)

The simplest device for measuring the energy distribution of electrons is the retarding field analyzer. In its simplest form it consists of an optically highly transparent grid between the electron emission source and the collector. A negative potential U_r is applied between the grid and the source so that all electrons with kinetic energies less than U_r are repelled and do not arrive at the collector. U_r is the effective potential, that is the voltage difference between the Fermi levels of the sample and grid plus the difference between the work functions of both materials. In practice it is necessary to establish a field free region between the sample and the analyzer, which is accomplished by placing a second grid held at sample potential between the repeller grid and the sample. The optimum position for this grid is midway between the sample and the analyzer [21]. The energy resolution of the system can be further improved by replacing the single repeller grid with a double grid which reduces field inhomogeneities caused by the finite size of the mesh. A fourth grid is frequently mounted between repeller and collector to reduce the capacity between them and prevent any ac modulating voltage applied to the repeller from being passed on to the collector directly. The grids and collector are best constructed as concentric hemispheres with the sample at the centre to give uniform trajectories to all electrons emitted from the target. The arrangement which is shown schematically in fig. 1.4

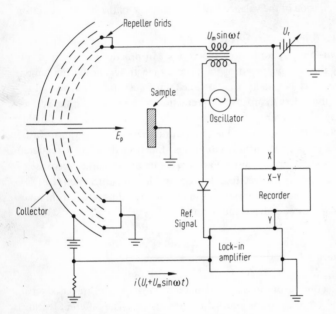

Fig. 1.4. Circuit for using a 4 grid LEED optics as an retarding field analyzer (RFA).

can also be used as detector for LEED experiments (although for this only the first two of the four grids are strictly necessary) by coating the collector with a fluorescent screen, and changed potentials applied to the analyzer. When used as an energy analyzer the collector is normally biased some hundred volts positive to aid collection of slow electrons emerging from the grid section and to inhibit secondary emission from the collector itself. Local charging of the phosphor on the collector surface also becomes unimportant. The negative potential applied to the repeller grids is varied continuously between $-U_{max}$ and zero. The collector current $i(U_r)$ as a function of the (effective) retarding potential U_r is then given by

$$i(U_r) \propto \int_{E=eU_r}^{\infty} N(E)\,\mathrm{d}E \tag{1.5}$$

where $N(E)$ is the energy distribution of secondary electrons emitted from the sample, assuming that secondary electrons created at the grids may be neglected. The retarding field analyzer separates those electrons with energies less than eU_r from the rest which are all collected. Consequently, peaks in the $N(E)$ distribution appear as small steps on a slowly varying background in the $i(U_r)$ curves. The energy distribution N (E) is obtained by differentiation:

$$\frac{\mathrm{d}\,i\,(U_r)}{\mathrm{d}\,U_r} \propto N(E) \tag{1.6}$$

This operation is performed electronically for example as proposed by Leder and Simpson [22]. An ac voltage $U_m \cdot \sin \omega t$ with a small amplitude U_m is superimposed on the retarding potential U_r. The collector current $i(U_r + U_m \sin \omega t)$ may then be expressed in the form of a Taylor expansion:

$$i(U_r + U_m \sin \omega t) = i(U_r) + i'(U_r) \cdot U_m \sin \omega t + \frac{i''(U_r)}{2!} \cdot U_m^2 \sin^2 \omega t + \ldots \tag{1.7}$$

The prime denotes differentiation with respect to U_r. The amplitude of the signal appearing with frequency ω is given by

$$A(\omega) = U_m \cdot i'(U_r) + \frac{U_m^3}{8} \cdot i'''(U_r) + \ldots \tag{1.8}$$

If U_m is small enough (in practice of the order of 1 V_{rms}) the amplitude of the component of the collector current with frequency ω is proportional to $i'(U_r)$ and hence proportional to $N(E)$. This component can be selected using a phase-sensitive detector or lock-in-amplifier. A circuit suitable for use with a four grid optics is shown in fig. 1.4. The modulating ac voltage is coupled into the retarding voltage through a transformer. The capacitance between collector and repeller grids can be completely compensated using a bridge circuit. A further improvement is possible by including a simple preamplifier in the collector circuit [23].

The Auger spectra usually take the form of small, relatively sharp peaks on a slowly varying background which can be considerably reduced by a second differentiation. This is as readily achieved electronically as the first differentiation since the Taylor expansion of $i(U_r + U_m \sin \omega t)$ also contains a component with frequency 2ω whose amplitude is given by

$$A(2\omega) = \frac{U_m^2}{4} \cdot i''(U_r) + \frac{U_m^4}{48} \cdot i^{(4)}(U_r) + \ldots \tag{1.9}$$

If U_m is not too large the desired second derivative of $i(U_r)$ is proportional to $A(2\omega)$ and also in turn to $\dfrac{\mathrm{d}\,N(E)}{\mathrm{d}\,E}$. Experimentally $A(2\omega)$ is recorded by tuning the phase-sensitive detector to 2ω.

A further advantage of measuring $i''(U_r) \propto \dfrac{\mathrm{d}\,N(E)}{\mathrm{d}\,E}$ is that the capacitively coupled current has no 2ω component and is therefore not registered by the detector. Furthermore, many of the peaks in the $N(E)$ curve have a tail on the low energy side, so that in practice the position of the differentiated peak on the energy scale is more easily defined.

The resolution of a retarding field analyzer is limited by instrumental factors and by the amplitude of the modulating ac voltage. The instrumental resolution is about 1%; which means at 100 eV electron energy two peaks may be detected if their energies differ at least by 1 eV. The amplitude of the modulating ac voltage is usually between 1 and 10 eV and this parameter usually determines the total resolution but since it also determines the sensitivity of the method it is normally undesirable to reduce the modulation significantly. The limit of sensitivity is given by the strength of the signal current compared with the magnitude of the noise current, the signal to noise ratio. The largest single noise component is shot noise which is thermionic in origin so that its current becomes $i_N = \sqrt{2ei \cdot B}$, where i is the collector current and B the bandwidth.

The disadvantage of the retarding field analyzer is that the collector current is composed of all secondary electrons with energies exceeding $-eU_r$, whereas information is required only from a small range within eU_m around eU_r. The collector current i is therefore much greater than it would be if only the electrons of interest were collected. The resulting shot noise must be counteracted by decreasing the bandwidth of the amplifier which in turn increases the time required per spectrum. Under typical operation condition $B \approx 0.05$ Hz, in other words the amplifier time constant is several seconds so that a full spectrum takes perhaps 20 min. to record. Furthermore, the primary current must be relatively large (≥ 10 μA) in order to obtain enough signal, which increases the probability of surface changes caused by electron bombardment.

On the other hand the angle of acceptance of RFA is comparatively large, some 120°, so that the currents measured are large enough for straightforward electronic techniques to be used. Using higher derivatives of the energy distribution can help in resolving fine structure. As mentioned earlier the four grid optics represent an extremely convenient method for displaying LEED patterns which is the reason why this arrangement has been the most widely used surface research tool for the last few years.

1.5.2. Cylindrical Mirror Analyzer (CMA)

The problem of the excessively high detector currents experienced with retarding field analyzers can be avoided by using a dispersion type analyzer, in which an image of the source or entrance slit is projected onto the collector. The electrons are focussed either electrostatically or magnetically in such a way that only those with energies within a certain range, usually selected by an exit slit, form the image. Thus the collector current consists only of electrons which had energies within the pass band of the analyzer.

One of the most important analyzers of this type is the cylindrical mirror analyzer, originally developed by Blauth [24] and optimized theoretically by Hafner et al. [25] and Sar-el [26]. The analyzer is shown schematically in fig. 1.5. A potential U_a is applied between the two coaxial cylindrical electrodes which creates an electrical field with cylindrical symmetry. The outer cylinder is held at a negative potential with respect to the inner cylinder which is usually grounded. Electrons entering the analyzer through the annular entrance are deflected towards the inner cylinder by an amount depending on their initial kinetic energy. If a potential U_a is applied to the outer cylinder, electrons with energy $E_e = eU_e$ pass through the exit slit onto the collector, which is for example the first dynode of an electron multiplier. The ratio U_e/U_a will be between 1.5 and 2 for an analyzer of the usual diameter and length. The ratio remains constant over a very wide range of energies. The energy resolution $\Delta E/E$ is constant and the current at the exit aperture is given by $i \propto N(E) \cdot \Delta E$, or since $\Delta E \propto E$,

$$i \propto E \cdot N(E) \tag{1.10}$$

Again the modulation techniques as described for the retarding field analyzer may be applied.

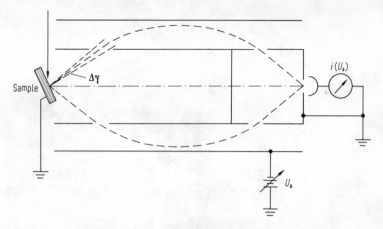

Fig. 1.5. Schematic cross-section of the cylindrical mirror analyzer (CMA).

The amplitude $A(\omega)$ of the signal at frequency ω is then given by

$$A(\omega) \approx U_m \cdot i'(E) \propto \frac{\mathrm{d}N(E)}{\mathrm{d}E} \qquad (1.11)$$

In contrast to the RFA the *first* derivative of the collector current yields $\mathrm{d}N(E)/\mathrm{d}E$ and not the second derivative.

The cylindrical mirror analyzer is superior to other analyzers of the dispersion type owing to its second-order focussing properties for an entrance angle $\gamma = 42°$. Electrons passing through the entrance slit with an angular spread of $\Delta\gamma$ are focussed to a ring with radius proportional to $(\Delta\gamma)^3$ about the analyzer axis. The effect of this is that electrons are accepted over a considerable solid angle for a given energy resolution so that the transmission of this type of analyzer is exceptionally high. The luminosity (product of solid angle of acceptance and effective source size) reaches about half the value of the RFA. The major advantage of the CMA over the RFA is the big reduction in shot noise. Only electrons within a small energy range ΔE are transmitted to the collector, which in practical terms represents a collector current reduction by a factor of 10^4 over that of RFA. Since the shot noise current is given by $i_N = \sqrt{eBi}$, the actual improvement in signal to noise ratio is a factor of 100, which allows for example, the bandwidth B to be increased so that faster scanning times are possible, or much smaller primary beam currents to be used to minimize electron bombardment effects on the surface.

The instrumental resolution of the CMA is in the first instance governed by the width of the entrance and exit slits, but may also be influenced by stray magnetic fields – especially in the low energy range. Commercial instruments are normally designed to a resolution similar or slightly better than that of a four-grid LEED optics.

1.5.3. 127°-analyzer

The 127° analyzer as proposed by Hughes and Rojanski [27] is also an analyzer of the dispersion type using an electrostatic field. The principle is indicated in fig. 1.6. Electrons entering the

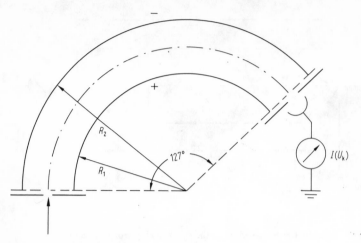

Fig. 1.6. Schematic cross section of the 127° analyzer.

cylindrical section through the entrance slit are deflected by the transverse electric field created by the voltage U_k applied between the capacitor plates. An electron with velocity v describes a circular path of radius R, depending on U_k. To a first approximation electrons with energy $E_e = e\,U_e$ strike the exit slit if the following relation between U_e, U_k and the radii of curvature R_1 and R_2 is fulfilled:

$$\tfrac{1}{2} U_k = U_e \ln (R_1/R_2) \tag{1.12}$$

At fixed values of R_1 and R_2 the electron energy distribution is measured by varying U_k. Again the resolution is mainly determined by the width of the entrance and exit slits. A sector of 127° yields optimum directional focussing. Using a highly sensitive electron multiplier as detector it is possible to use very narrow slits providing very good instrumental resolution. This type of analyzer can be made quite small and is therefore suitable for use as a movable device for example for studying the angular dependence of emitted electron distributions [28].

1.5.4. Concentric hemisphere analyzer (CHA)

The concentric hemisphere analyzer is another type which uses electrostatic deflection. It was initially proposed by Purcell [29], further improved by Meyer et al. [30] and by Kuyatt and Simpson [31], and developed to a very high performance by Siegbahn [32] for use in photo-electron spectroscopy (ESCA). The principle of this analyzer is illustrated by fig. 1.7.

Electrons with energy $E_e = e\,U_e$ passing through the entrance slit are focussed onto the exit slit if the following relation between the potential difference U_k across the two hemispheres, their radii R_1 and R_2, and U_e is fulfilled:

$$U_k = U_e\,(R_1/R_2 - R_2/R_1) \tag{1.13}$$

The CHA is double focussing, i.e. it focusses in two planes. This analyzer type is normally used in x-ray photoelectron spectroscopy (ESCA), together with a pre-retarding device which

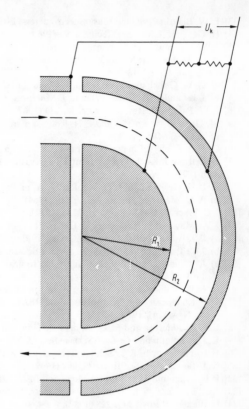

Fig. 1.7. Schematic cross section of the concentric hemisphere analyzer.

is a simple technique for increasing the resolution at higher electron energies [33]. The resolution of a dispersion analyzer is a fixed percentage, say 0.5 %, of the transmission energy. The effective separation of electron energies is therefore better at small transmission energies. If electrons with initial energies around 1 000 eV are decelerated by a variable retarding potential between the sample and the entrance slit of the analyzer down to 100 eV, then an analyzer resolution of 0.5 % enables the separation of lines with an energy difference of only 500 meV, whereas without retardation the best resolution would have been 5 V. In other words, this represents an improvement in the effective resolving power of the spectrometer of an order of magnitude. The retardation technique is not restricted to hemispherical analyzers; CM analyzers with retardation grids are also commercially available.

Retardation is primarily used in ESCA spectrometers where, owing to the relatively high energies of available x-ray sources, analyzer resolution is at a premium.

1.6. References

[1] J. W. May, Adv. Catalysis *21*, 151 (1970).
[2] a) P. A. Redhead, J. P. Hobson and E. V. Kornelsen: The Physical Basis of Ultrahigh Vacuum, Chapman and Hall, London 1968.
   b) N. W. Robinson: The Physical Principles of Ultrahigh Vacuum Systems and Equipment. Chapman and Hall, London 1968.
   c) E. A. Trendelenburg: Ultrahochvakuum. G. Braun, Karlsruhe 1963.

[3] E.W. Blauth: Dynamic Mass Spectrometers. Elsevier, Amsterdam 1966.

[4] H. Gentsch, J. Vac. Sci. Techn. *6*, 93 (1969).

[5] J.W. May and L.H. Germer, J. Chem. Phys. *44*, 2895 (1966).

[6] R.M. Mobley, in: Methods of Experimental Physics, (L. Marten, ed.). Academic Press, New York 1967, Vol. 4B, p. 318.

[7] a) G.W. Gobeli and F.G. Allen, J. Chem. Phys. Solids *14*, 23 (1960).

   b) J.J. Lander, G.W. Gobeli and J. Morrison, J. Appl. Phys. *34*, 2298 (1963).

[8] P.W. Palmberg, Rev. Sci. Instr. *38*, 834 (1967).

[9] M. Green, J.A. Kafalas and P.H. Robinson, in Semiconductor Surface Physics. (R.H. Kingston ed.) Philadelphia 1957, p. 349.

[10] G. Ertl, Z. phys. Chem. N. F. (Frankfurt) *50*, 46 (1966).

[11] G. Ertl and J. Koch, Proc. Vth Int. Congress on Catalysis, Palm Beach 1972, North Holland, Amsterdam 1973, p. 969.

[12] a) L.H. Germer and A.U. MacRae, J. Appl. Phys. *33*, 2923 (1963).

   b) A.U. MacRae, Surface Sci. *1*, 319 (1964).

[13] J.C. Tracy, J. Chem. Phys. *56*, 2736 (1972).

[14] J.W. May and L.H. Germer, J. Chem. Phys. *44*, 2895 (1966).

[15] a) P.J. Estrup and J. Anderson, J. Chem. Phys. *45*, 2254 (1966).

   b) J.C. Tracy and J.M. Blakely, Surface Sci. *15*, 257 (1969).

[16] a) R. Gomer: Field Emission and Field Ionization. Harvard University Press, Cambridge (Mass.) 1961.

   b) E.W. Müller and T.T. Tsong: Field Ion Microscopy. Elsevier, New York 1969.

[17] D.F. Klemperer, in: Chemisorption and Reactions on Metallic Films. (J.R. Anderson ed.), Academic Press, New York 1971.

[18] L.T. Chadderton and M.G. Anderson, Thin Solid Films *1*, 229 (1968).

[19] K. Christmann and G. Ertl, Surface Sci. *33*, 254 (1972).

[20] H.E. Farnsworth, R.E. Schlier, T.H. George and R.M. Burger, J. Appl. Phys. *29*, 1150 (1958).

[21] N.J. Taylor, Rev. Sci. Instr. *40*, 462 (1969).

[22] L.B. Leder and J.A. Simpson, Rev. Sci. Instr. *29*, 571 (1958).

[23] J.C. Tracy and G.K. Bohn, Rev. Sci. Instr. *41*, 591 (1970).

[24] E. Blauth, Z. Phys. *147*, 228 (1957).

[25] H. Hafner, J.A. Simpson and E.C. Kuyatt, Rev. Sci. Instr. *39*, 33 (1968).

[26] H.Z. Sar-el, Rev. Sci. Instr. *38*, 1210 (1967).

[27] A.L. Hughes and V. Rojansky, Phys. Rev. *34*, 284 (1929).

[28] B. Feuerbacher and B. Fitton, Verh. Dt. Phys. Ges. (VI) *8*, 444 (1973).

[29] E.M. Purcell, Phys. Rev. *54*, 819 (1938).

[30] U.D. Meyer, A. Skerbale and J. Lassettre, J. Chem. Phys. *43*, 805 (1965).

[31] C.E. Kuyatt and J.A. Simpson, Rev. Sci. Instr. *38*, 103 (1967).

[32] K. Siegbahn, C. Nordling, A. Fahlmann, R. Nordberg, K. Hamrin, J. Hedman, G. Johansson, T. Bergmark, S.E. Karlsson, J. Lindgren and B. Lindberg: ESCA-Atomic, Molecular and Solid State Structure studied by means of Electron Spectroscopy. Uppsala 1967.

[33] J.C. Helmer and N.H. Weichert, Appl. Phys. Lett. *13*, 266 (1968).

[34] C.R. Brundle, J. Vac. Sci. Techn. *11*, 212 (1974).

2. Auger electron spectroscopy

2.1. Historical development

In 1925 P. Auger [1] observed traces of electrons on photoplates which had been exposed to hard x-ray radiation. He interpreted the formation of these electrons as a radiationless transition in atoms excited by the primary x-ray photons, whereby the emission of an electron from an outer level with an energy equal to that released by the filling of the core hole competes with the emission of characteristic x-ray radiation. These "Auger"-electrons have since been the subject of extensive theoretical [2, 3] and experimental [4, 5] investigations.

In 1953 Lander [6] studied the energy distribution of secondary electrons emitted from solid samples irradiated with slow electrons. He observed small characteristic peaks which he attributed to Auger electrons. He also pointed out that these electrons might be used as a tool for surface analysis, but it was not until 1967 that the idea was followed up.

Tharp and Scheibner [7] demonstrated that the ordinary electron optics for LEED experiments could be used as a retarding field energy analyzer for the detection of Auger electrons by differentiation of the integral energy distribution. A disadvantage however was the fact that only very small peaks appeared on the large slowly varying background of secondary electrons which are emitted from a surface bombarded by low energy primary electrons. Under these conditions it is rather difficult to identify the transitions.

An important step forward was made by Harris [8, 9] who used a 127° energy analyzer and registered the electronically differentiated energy distribution $dN(E)/dE$ as suggested by Leder and Simpson [10]. Almost simultaneously Weber and Peria [11] and Palmberg and Rhodin [12] used electronic differentiation techniques in combination with conventional 3-grid LEED electron optics and obtained spectra comparable in quality to those of Harris.

Further progress was made by Palmberg [13], who showed that the addition of a fourth grid to a commercial 3-grid LEED optics improved the resolution of the spectrometer considerably. Furthermore, a marked increase in the sensitivity could be obtained by using a glancing angle primary electron gun instead of the usual excitation by electrons striking the surface under normal incidence.

Another important step in the experimental development of this method was made by Palmberg et al. [14] by using a cylindrical mirror analyzer to detect Auger electrons. With this technique the sensitivity, resolution and signal to noise ratio could be improved to the extent that even a fast oscillograph scan of the Auger spectra and the detection of surface impurities with concentrations as low as 10^{-3} of a monolayer became possible.

Nowadays the trend is to minimize the diameter of the primary beam in order to increase the lateral resolution of the surface analysis, which may also be combined with scanning the primary beam across a portion of the surface to localize the source of the spectra.

Although Auger electron spectroscopy was introduced into surface research only six years ago, it is at present perhaps the most important tool. More than several hundreds of papers have been published since then, which are quoted in bibliographies of Haas et al. (1971) [15], and Hawkins (1972) [16]. Some critical surveys and review papers are contained in the reference list [17]–[22], [90, 91].

2.2. Experimental

The essential parts of an Auger electron spectrometer are, as is usual for spectroscopic techniques, a source for primary excitation, the sample, and an analyzer and detector system.

2.2.1. The source of excitation

In principle any type of radiation which is able to ionize the inner shells of atoms can be used to excite Auger electrons. The most convenient means is an electron beam, which has now become more or less the standard source. A straightforward electron gun with focussing and deflection electrodes readily provides a beam of the required intensity. Since the energy of these primary electrons does not need to be homogeneous, no energy filtering is necessary. Commercial electron guns produce emission currents up to 200 μA with energies of usually 2–3 keV. Focussing onto a spot of about 0.2×0.2 mm^2 presents no problem. More recently electron guns with spot diameters as small as 25 μm and less have been developed for AES work, allowing the analysis of microscopic surface regions. Further improvements are to be expected in this direction. Replacing the hot cathode by a field emission source should be advantageous, enabling the reduction of the size of the primary spot to a few 100 Å at which point a true microanalysis becomes possible.

Since the atomic ionization cross sections for electron impact are in the range of 10^{-20} cm^2 and are fairly independent of the primary energy, this type of excitation lends high sensitivity to AES. A disadvantage is that the Auger peaks are on a relatively large background which has to be reduced by the differentiation technique already mentioned. Secondary ionization by back-scattered electrons also influences the Auger current and thereby complicates quantitative interpretation.

A further important point is the dependence of the Auger emission on the angle of incidence of the primary electrons. As shown by Palmberg [13] the Auger yield (and therefore also the sensitivity for detection) from the topmost atomic layer is strongly enhanced relative to that of the substrate if the angle between the primary beam and the surface is decreased and approaches grazing incidence. Qualitatively this effect can be readily understood, since at grazing incidence the primary beam penetrates less into the substrate and excites more particles of the surface layer. At very small angles ($<6°$) the total Auger yield decreases because of surface roughness. Best results are therefore obtained with an angle of incidence of about 10–15°. In the case of Fe evaporated onto Si(111) an enhancement of the Fe Auger signal height by a factor of 10 compared with that under normal incidence of the primary beam has been reported [23].

Auger electrons may also be excited by x-ray radiation as is the case with ESCA work. Maintaining an x-ray source in ultra high vacuum is more complicated than an electron gun, and the time needed for recording a spectrum is much longer. This is mainly due to the fact that with normal x-ray tubes maximum photon fluxes at the sample of about 10^6 photons/s are attained [92] whereas a primary electron current of 1 μA corresponds to an electron flux of $\sim 6 \cdot 10^{12}$ electrons/s. This effect is partly compensated by the fact that photoionization cross sections may be several orders of magnitude larger than those for electron impact ionization. Moreover the background is much lower with excitation by means of x-rays since the large contribution from the "rediffused" primary electron background is absent. Spectra are therefore recorded as $N(E)$ instead of dN/dE. Furthermore the damage caused by the primary radiation on sensitive surface layers is much less than with an electron beam. Examples of x-ray induced Auger spectra are given by Siegbahn et al. [24], and by Smith and Huchital [25].

Finally it should be mentioned that excitation may also be achieved by high energetic particles, as demonstrated by Musket and Bauer [26] using 350 keV protons. Though this technique has some advantages (in particular for quantitative analysis) it will certainly be restricted to only few applications.

2.2.2. The sample

In principle any solid, single- or polycrystalline, can be used as a sample. A smooth flat surface improves the quality of the spectra, but is not essential. It is very important that the sample should be correctly positioned with respect to the analyzer. For this reason the sample has to be mounted on a manipulator of the type described in more detail in the LEED section. Charging effects may be avoided by grounding the sample. If it has poor electrical conductance (insulators or semiconductors) it is necessary to produce enough charge carriers to compensate the excess charge. In most cases this is automatically achieved by the high energy primary electrons which eject enough secondary electrons. In difficult cases the use of a glancing incidence beam may increase the secondary yield to the necessary level. An evaporated metal film can also be used to conduct excess charge to ground.

It is also possible to record Auger spectra from samples at elevated temperatures. Heating is possible for example by radiation from a filament behind the sample, by laser light or by conduction from a resistive heater.

2.2.3. Analyzer and detector system

A 127° energy analyzer was used in the pioneering work of Harris [8], [9]. The arrangement of this type of spectrometer is shown in fig. 2.1. The primary electron beam strikes the sample

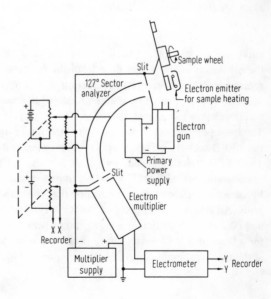

Fig. 2.1. Scheme of the apparatus used by Harris [8].

surface with an angle of about 15° and a small fraction of the created secondary electrons enters the entrance slit (width ~0.2 mm) of the analyzer. The signal is detected by an electron multiplier. Harris' spectrometer achieved the creditable resolution of 0.1 %, but a disadvantage of the method is the small transmission caused by the narrow entrance slit. This was compensated by using a multiplier which has however a different amplification for electrons of different energies leading to some discrimination of the spectra. This can be seen in the spectrum a) of fig. 2.2, where the decrease of $N(E)$ at lower energies is caused by this effect. Harris then

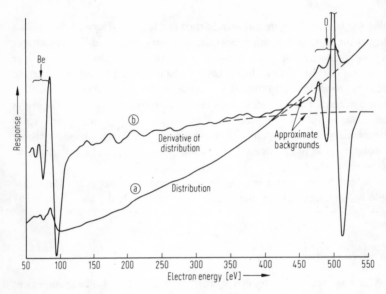

Fig. 2.2. Auger spectra from a beryllium sample [8].
a) Energy distribution $N(E)$. b) First derivative $\mathrm{d}N(E)/\mathrm{d}E$.

applied electronic differentiation technique in order to obtain $\mathrm{d}N(E)/\mathrm{d}E$, which facilitates the extraction of the Auger data from the background as is clear from curve b. It must be remembered, however, that the evaluation of the intensity of the Auger emission (for quantitative analysis) is complicated by the amplitude of the modulation voltage used for differentiation and the change in the peak shape.

The most popular analyzer at present is the retarding field grid analyzer in combination with the LEED technique [11], [12]. The experimental device is shown schematically in fig. 2.3. The LEED technique needs only 2 or 3 grids, but a 4 grid optics improves the resolution considerably for AES [13]. A further improvement of the signal/noise ratio can be obtained by using a capacitor bridge to neutralize the capacitance between the collector and the retarding grids. An Auger spectrometer of the RFA type can be used to record a spectrum between 0 and 1 000 eV in about 10 min using a primary beam current of about 100 μA. Further refinements for a three grid RFA system have been proposed by Skinner and Willis [27] which involves in the main cross modulation of electron gun and analyzer grids for suppression of unwanted signal.

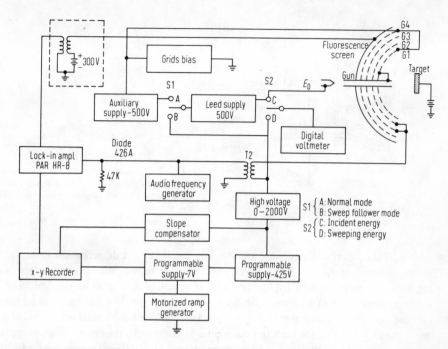

Fig. 2.3. Schematic diagram of the LEED-Auger device as described by Chang [18]. The sweep follower mode eliminates all energy loss peaks from the spectrum.

The LEED optics is equipped with an electron gun usually providing electrons at incidence normal to the surface. As outlined in the preceding section the surface sensitivity is increased by using a large angle of incidence [13], so that a second electron gun is often used. The lower limit of detection of surface species is then in the range of about 1% of a monolayer (e.g. of O or C).

The essential disadvantages of the retarding field analyzer are the relatively long times needed for recording a spectrum and the high currents of the primary beam. It is thus not possible to study kinetic processes which occur in times of less than a few minutes. A typical primary electron beam current for use with an RFA would be about 100 μA. With a primary energy of 3 kV and a spot area of ~ 1 mm^2 this corresponds to a power of about 30 W/cm^2 leading to considerable local heating effects. The temperature of a metal disc of 1 mm thickness and 5 mm diameter increases by about 60 °C; the surface temperature at the point of impact is certainly even higher. Parallel to this heating effect adsorbed layers may also be strongly influenced by electron beam stimulated desorption or dissociation. Both effects are minimized by the use of the cylindrical mirror analyzer (CMA).

The cylindrical mirror analyzer was introduced to AES by Palmberg et al. [14] and is at present the best instrument for this method on grounds of performance and construction. The arrangement of this analyzer in an Auger spectrometer is shown schematically in fig. 2.4. The target may, as in the case of the RFA, be bombarded by electrons under glancing or normal incidence. In the latter case the electron gun is mounted inside the analyzer. The advantage of this arrangement is the simpler adjustment of the sample position.

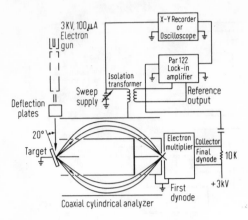

Fig. 2.4. Schematic diagram of an Auger device using the cylindrical mirror analyzer and a glancing angle primary electron gun. After Palmberg et al. [14].

Also significant is that when using a CMA and fast harmonic detection the measured signal is proportional to $E \cdot dN(E)/dE + N(E)$. At energies below 200 eV the two terms are of opposite sign resulting in a flat background. If a multiplier is used as the first stage of the detector system its gain decreases with decreasing energy. On the other hand Auger intensities increase considerably with decreasing energy. These effects cancel each other so that a whole spectrum may be recorded with the same gain, in contrast to the RFA. In some systems the first dynode of the multiplier can be given additional bias voltage to compensate for the loss in sensitivity for low energies. With the CMA the signal/noise ratio is improved by a factor ~ 100 compared to the RFA. Using similar primary electron currents this leads to an enormous increase in recording speed: A complete spectrum between 0 and 1 000 eV may be recorded fast enough to be displayed on an oscilloscope. Alternatively, using a larger time constant the primary beam current may be reduced to less than 0.1 μA thereby minimizing the heating and damage caused by the electron beam. The ultimate sensitivity of the CMA is better than 0.1% of a monolayer.

2.2.4. Further refinements

A series of further developments and improvements of existing experimental devices has been realized.

The energy distribution of secondary electrons contains besides the Auger electron transitions further peaks caused by plasmon excitations, inelastic losses etc. The energy of the Auger electrons does not depend on the energy of the primary electrons E_p in contrast to the characteristic losses whose position on the energy scale corresponds to a fixed distance to E_p, It is therefore possible to separate both parts by suitable changes in the electronics as proposed by Gerlach et al. [28]. If the potential of the sample is modulated and the primary energy E_p and the transmission energy E of the analyzer are fixed, then the energies of the Auger electrons with respect to E are modulated and the Auger electrons are detected but not the characteristic losses. The latter may be recorded if just the primary energy E_p is modulated.

With the RFA a large background appears in the spectra which is characteristic of the general shape of the secondary electron distribution function. This background may be removed by

injecting electronically the mirror image of the secondary electron emission at the output of
the RFA [29]. A similar procedure was proposed by Fiermans and Vennik [30] who reduce the
noise of $N(E)$ curves by using a signal averager and digital techniques and subtract a suitable
function as the background contribution. The latter is however not free from problems.

The effective scanning time of the RFA may be shortened by using a multichannel monitor
as developed by Sickafus and Colvin [31]. The technique consists simply in omitting unwanted
portions of the spectrum by changing the scan rate. The peak heights of the signals from a
series of selected energies (= elements) are recorded continuously. This method is suitable for
following the kinetics of surface processes. However a fast total scan time for the whole spectrum
is no serious problem if a cylindrical mirror analyzer is used.

A very promising refinement of the method is to observe the lateral distribution of elements on
a surface. This is very useful in many problems in metallurgy, semiconductor technology,
catalysis etc. and can be achieved by scanning the primary beam across the sample surface
with simultaneous monitoring of the Auger amplitude of a chosen element [32]. Scanning
may be achieved by continuous deflection of the primary beam or (less conveniently) by moving
the sample. Fig. 2.5 shows as an example the distribution of sulfur on a Pt surface after partial

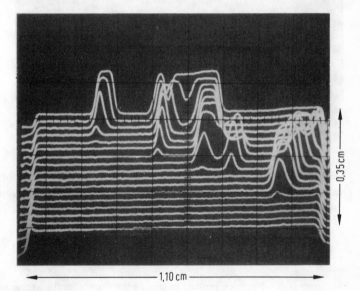

Fig. 2.5. Sulfur distribution
on a Pt(110) surface after
partial reaction with oxygen as monitored by scan-
ning the primary electron
beam across the sample
surface.
After Bonzel [33].

0,35 cm

1,10 cm

reaction with oxygen [33]. The partial resolution is not very good in this example but may be
further improved by decreasing the diameter of the primary beam.

A currently available commercial system [34] offers a lateral resolution of about 20 μm.
With this instrument a magnified image of the surface is displayed on a monitor screen using
the total current collected by the sample which is modified by the local variations of the secondary
emission coefficient on the surface. As an example the image from an integrated circuit is
shown in fig. 2.6. The purpose of the display is to enable the area on the surface which is to
be analyzed to be identified and returned to repeatedly. It is to be expected that progress in
this direction will be rapid, steady increases in the lateral resolution and refinements such as
display of Auger electron images being prospects for the immediate future.

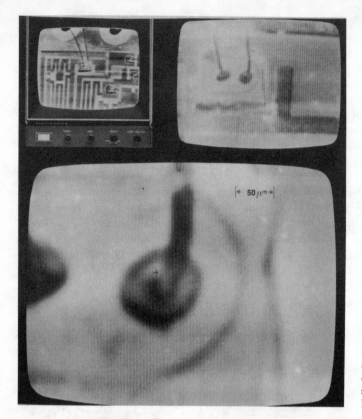

Fig. 2.6. Image from a part of an integrated circuit by scanning the primary electron beam and monitoring the current through the sample. This device enables also the exact positioning of the area which is analysed by AES. (Varian Corp., Palo Alto).

One approach under investigation in several laboratories consists of combining AES with a scanning electron microscope. This was first achieved by MacDonald and Waldtrop [35] using a cylindrical mirror analyzer, but under poor vacuum conditions.

2.3. Mechanism of the Auger process

An atom which has been ionized in one of the inner (core) states may return to its electronic ground state via one of the following processes:

a) An electron of an energetically higher level "jumps" into the core hole, the energy thereby released is emitted as a quantum of characteristic x-radiation (Fig. 2.7a).

b) The core hole is filled by an outer electron, but the available energy is transmitted in a radiationless process to a second electron which may then leave the atom with a characteristic kinetic energy (Fig. 2.7b).

This second process is called the Auger effect. A simple explanation could be in terms of an internal photoeffect i. e. in the primary process an x-ray photon is formed which excites a

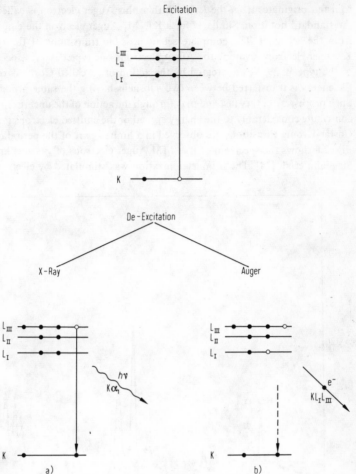

Fig. 2.7. Processes for de-excitation of atomic core holes. a) Emission of x-ray radiation. b) Emission of an Auger electron.

photoelectron within the atom and is thereby destroyed. This interpretation is not correct, however, since the Auger effect is a true radiationless process without the intermediate formation of a "virtual" quantum of electromagnetic radiation. This is apparent from a comparison of the selection rules for optical processes with the observed electronic transitions for the Auger process. X-ray emission is governed by the selection rules for dipole radiation, which means that the quantum number of orbital angular momentum must change by ± 1. On the other hand the Auger process is governed by electrostatic forces produced by the interaction of a hole in an incomplete shell with its surrounding electron clouds and is not strictly dependent on selection rules.

Auger electrons are classified by referring to the energy levels involved in their production. The starting point is a core hole created by ionization of the atom. The means of doing this is unimportant; identical processes are observed after excitation with electrons or x-rays as well as with nuclear conversion and K-capture. This hole – for example in the K shell – may be filled by transition from a higher shell, say L_2. If the emitted electron, excited by energy transfer, originates from the L_3 shell, then this Auger electron is called a KL_3L_3 electron in the standard notation. Similarly for a $KL_2 M_1$-Auger electron the emitted electron originates from the M_1 shell. The complete set of possible transitions of the type KX_qY_r forms the K-Auger electrons. Among the Auger electrons one type deserves special mention. Electrons of the type $W_p W_q X_r$ are created by the action of so-called Coster-Kronig transitions where the energy is transferred between two subshells having the same principal quantum number. Such processes are very fast and may through the action of the uncertainty principle $\Delta E \cdot \Delta t > h$ contribute considerably to the energy spread of the emitted electrons ("lifetime broadening"). Coster-Kronig transitions are observed in a limited part of the periodic table.

Fig. 2.8 shows the spectrum of $L_{2,3}$ MM Auger electrons of gaseous krypton as measured by Siegbahn et al. [24]. The primary ionization was stimulated by electron impact.

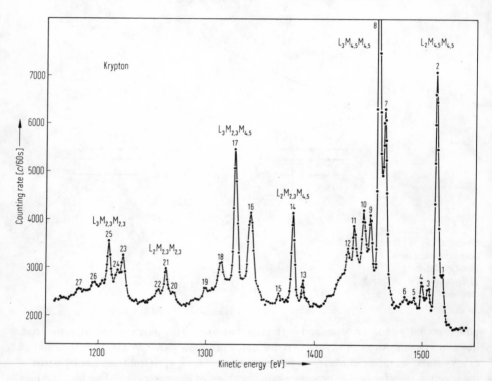

Fig. 2.8. Auger electron spectrum from gaseous krypton. The principal lines are indexed by the usual notation. (Note that the energy distribution $N(E)$ and not $dN(E)/dE$ is recorded). After Siegbahn et al. [24a].

The half width of the peaks in this spectrum is some 5–10 eV, – far more than the limiting instrumental resolution of about 0.1 eV. (The parameters determining the width of Auger peaks are discussed below.) In contrast to photoelectron spectroscopy the energy spread of the primary radiation is of no importance in AES, since the atom "forgets" the primary energy of excitation after ionization. This is mainly because the ionization process by electron impact occurs in less than 10^{-16} s, whereas the lifetime of a core hole is typically more than an order of magnitude larger.

There is a clear need to have a system for ordering the multiplicity of emission lines shown in fig. 2.8 and furthermore to have at least a theoretical estimate of the numbers, energies and intensities of the possible Auger transitions for the various elements. This must be based on some knowledge of the transition mechanism involved.

The energy level scheme of fig. 2.9 shows the correlation between the notation for Auger

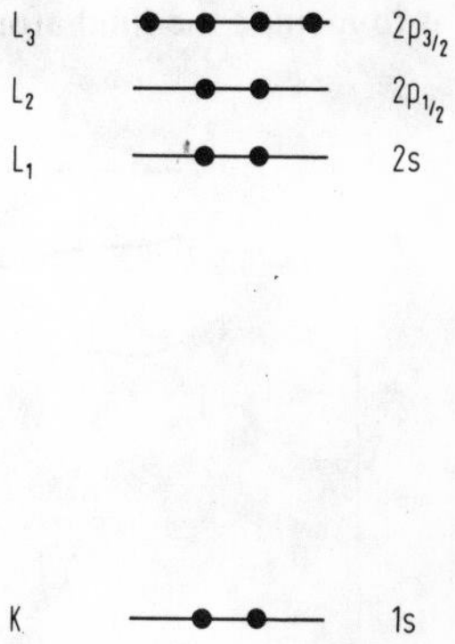

Fig. 2.9. Energy term scheme for atomic K and L levels.

transitions and the quantum number concept of x-ray spectroscopy. Following this picture six different types of KLL transitions should occur: KL_1L_1, KL_1L_2, KL_1L_3, KL_2L_2, KL_2L_3 and KL_3L_3. But sometimes *nine* different transitions of the KLL type (that means primary ionization of the K shell and final state with two vacancies in the L shell) are observed experimentally. The reason is that the x-ray level notation holds strictly only if strong spin-orbit interactions for the electrons take place, i.e. in the case of j-j coupling. However for the energy range of interest in AES (less than a few keV) the Coulomb interaction between the electrons is much stronger than the spin-orbit interaction thus leading to L-S (or Russel-Saunders) coupling, so that two final states $2s^1 p^5$ are possible for the doubly ionized atom after a KL_1L_2 transition:

The final states for all types of KLL transitions are listed in table 2.1. The whole picture is

Table 2.1. Final states and electronic configurations for all types of KLL transitions.

Transition	Final state	Electronic configuration
KL_1L_1	$^1S_0$	$2s^0\,2p^6$
KL_1L_2	$^1P_1, {}^3P_0$	$2s^1\,2p^5$
KL_1L_3	$^3P_1, {}^3P_2$	$2s^1\,2p^5$
KL_2L_2	$^1S_0$	$2s^2\,2p^4$
KL_2L_3	$^1D_2$	$2s^2\,2p^4$
KL_3L_3	$^3P_0, {}^3P_2$	$2s^2\,2p^4$

illustrated by fig. 2.10 where the relative energies of the different final states are plotted against the atomic number Z. In the case of pure LS-coupling ($Z \lesssim 20$) some of the different levels degenerate. In the intermediate range (intermediate coupling) nine different energies should occur, which has been observed for example in the Auger spectra of Zr by Hörnfeldt et al. [4]. At high atomic numbers ($Z \gtrsim 80$) the number of transitions is again reduced due to the operation of j-j-coupling. In all cases where no strong j-j-coupling occurs the usual convention is to include the final atomic state in the notation of the transition, e.g. KL_2L_3 ($^1D_2$).

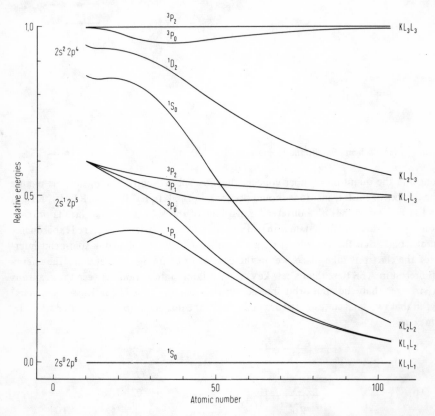

Fig. 2.10. Relative energies of KLL Auger transitions as a function of atomic number. The double ionized electronic configurations are given on the left. After Siegbahn et al. [24b].

In its use as a surface tool AES is normally restricted to energies below 1.5 keV. Since the energy of any particular type of transition (say KLL) increases rapidly with the atomic number, the types of transitions which are observed in the lower energy range depend on Z:

Primary excitation of shell	For elements with Z between
K	3 (Li) and 13 (Al)
L	11 (Na) and 35 (Br)
M	19 (K) and 70 (Yb)
N	39 (Y) and 94 (Pu)

This grouping follows directly from the binding energies of the electrons [24] and shows that the transitions of interest change from KLL to LMM, MNN, and NOO as surface studies of elements of increasing atomic number are made. A further change with increasing energy is that the Auger electron yields decrease considerably in favour of radiative transitions, and the ionization probabilities for the core electrons are also significantly reduced.

So far only Auger electrons emitted from isolated atoms have been discussed. With condensed matter a further complication arises from the fact that the valency levels exist as energy bands with more or less delocalized electronic states and finite energy width. For example with Si the LMM spectra are replaced by LVV transitions, since for this element the M states form the valence band. The process of Auger electron emission in such cases is illustrated schematically in fig. 2.11. After ionization of the L shell the hole is filled by an electron from the valence

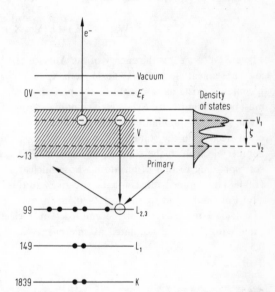

Fig. 2.11. Energy level diagram of Si illustrating an LVV Auger process. After Chang[18].

band and another electron is emitted from the valence band into the vacuum. Two further points have to be taken into consideration:

a) The initial kinetic energy of the electron is decreased by the work function of the solid material.

b) The line-shape of the energy distribution of the Auger electrons will be influenced by the energy distribution of the initial state, i.e. by the band structure. This topic is discussed below in some detail.

2.4. Energies and shapes of the Auger peaks

2.4.1. Free atoms

In many cases the Auger electrons are emitted from an outer shell having low binding energy so that to a first approximation their kinetic energy is equal to the binding energy of the deep lying level where the hole was initially created. A more exact treatment is of course much more complicated, since the emission of an Auger electron is influenced by many – electron effects in the inner shells. Quantum mechanical calculations for KLL transitions have been made for example by Asaad and Burhop [37] using certain assumptions about the ionization energies of K and L levels and the coupling between two vacancies in the L shell.

The energy of an Auger electron $W_o X_p Y_q$ is obviously given by

$$E(W_o X_p Y_q) = E(W_o) - E(X_p) - E(Y_q, X_p) \qquad (2.1)$$

where $E(W_o X_p Y_q)$ is the energy of the Auger electron which, after primary ionization of W_o and subsequent electron transition from X_p to W_o, is emitted from the state Y_q. (If no pure jj-coupling exists as discussed in the preceding section the different possible atomic states must be distinguished). $E(W_o)$ is the binding energy of an electron in the state W_o of the neutral atom, $E(X_p)$ is the binding energy of the electron in the level X_p, and $E(Y_q, X_p)$ the energy of an electron in the state Y_q, but moving in a potential of increased positive charge, since another electron in the state X_p is missing. This last energy term can be approximated as corresponding to that of an atom whose nuclear charge is ΔZ units larger. Bergström and Hill [38] determined values for the effective additional nuclear charge ΔZ for Hg experimentally: For the L_1 and L_2 shells $\Delta Z = 0.55$; for the L_3 shell $\Delta Z = 0.76$. It is therefore possible to calculate the energies of the emitted Auger electrons approximately for isolated atoms using equ. (2.1) and tabulated atomic energy level tables with estimates for ΔZ between 0.5 and 1.

2.4.2. Condensed matter

If the electronic states involved in a particular Auger process are deep lying core levels the situation with a solid sample will be quite similar to that with free atoms, since these states are not strongly affected by the chemical bond between the valence electrons. Only a correction for the work function which is usually of the order of a few eV has to be included in the interpretation.

If Auger electrons are emitted from the valence band of a solid (e. g. a metal) equ. (2.1) has to be modified. As can be seen from fig. 2.12 the Auger energy with respect to the vacuum level $E_{\text{vac, sample}}$ becomes

$$E(\text{SVV}) = E_s - E_1 - (E_2 + e\phi) \qquad (2.2)$$

where ϕ is the work function of the sample. When looking at the energy diagram of fig. 2.12 illustrating the principle of measurement it becomes evident that in fact the work function not of the sample but of the spectrometer ϕ_{sp} has to be taken into consideration. The energies

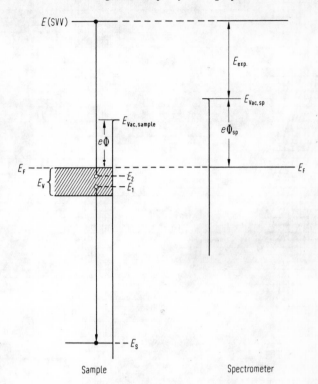

Fig. 2.12. Energy level diagram for
an AES experiment involving the
valence band of a solid.

E_1 and E_2 are located within the valence band of width E_v, which results in a maximum energy width of $2E_v$ for such Auger electrons. The probability for the participation of a certain energy level E depends on the electronic density of states $N(E)$, and therefore the line-shape of the Auger spectra is a function of $N(E)$. One would like to obtain information about the band structure from such measurements directly, but this is unfortunately rather complicated since the measured energy distributions are dependent on the transition probabilities.

Even more complex is the problem of calculating the energies of Auger electrons emitted from adsorbed layers. If all relevant processes take place in the core levels of the adsorbed particles then again equ. (2.1) with a correction for the work function will hold, but the valence states of substrate and adsorbed particles interfere mutually giving rise to new (chemisorption induced) energy levels, which introduce an extra and usually unknown factor.

For practical purposes however an approximate evaluation of the energies of the Auger transitions is sufficient. This can be achieved very easily under the assumption that the valence levels of adsorbed particles coincide with the Fermi level of the metal. Errors in the range of about 10 eV result which is normally quite acceptable bearing in mind the limited resolution and the (estimated) work function correction.

During the last few years a lot of experimental data on the Auger electron energies of the various elements has been obtained. A chart by Strausser and Uebbing [39] containing a collection of such data is reproduced in fig. 2.13. Since Auger spectra are usually recorded by the derivative technique as $dN(E)/dE$ curves the energy of an Auger transition is defined by convention as the minimum in the high energy wing of the differentiated peak [40].

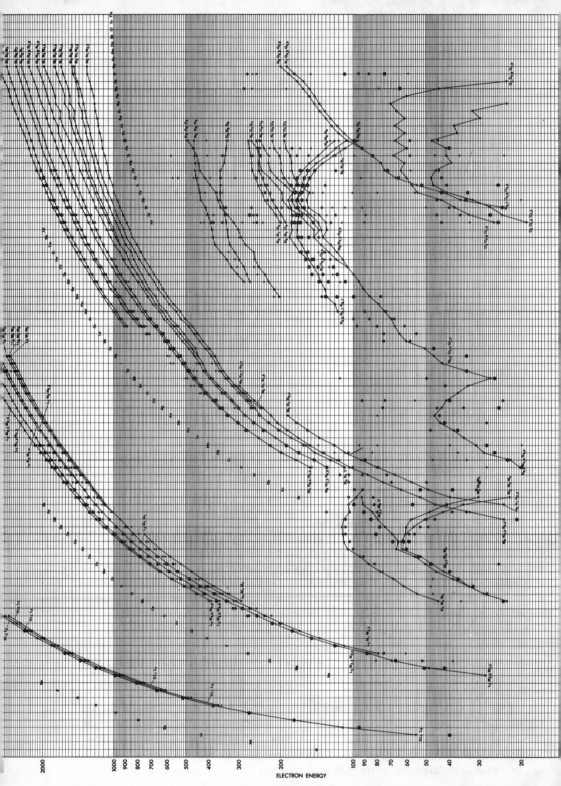

Fig. 2.13. Chart of Auger electron energies as a function of the atomic number. Strausser and Uebbing [39].

For an Auger electron emitted from an isolated atom the energy spread depends on the atomic levels involved and the lifetime of the excited state, provided the instrumental resolution is high enough. Lifetime broadening is caused by the uncertainty principle $\Delta E \cdot \Delta t > h$ and this is about 10 eV for the very fast Coster-Kronig processes. In fact these short lifetimes are determined experimentally from the energy broadening. Even shorter lifetimes are observed for plasmons; plasmon transitions may thus be distinguished from Auger peaks.

For Auger electrons originating from solids there is also a broadening of twice the width of the valence band and in addition from energy losses suffered by the emitted electrons. These losses (mostly interband transitions) usually give rise to a characteristic "tailing" at the low energy side of the non differentiated peak.

Besides these inherent effects the energy spread of the measured spectra is of course also affected by the limited resolution of the spectrometer and the amplitude of the ac modulation voltage. In particular the energies of the LMM transitions for the transition metals are separated only by small amounts, and therefore frequently the true line shape of an Auger peak is not observed but rather the superposition of several peaks.

2.4.3. Chemical effects

Chemical effects or the influence of the chemical environment on the properties of an atom cause three possible changes in Auger spectra:

a) The chemical bond ($\triangleq$ charge transfer) causes a shift of the energy of a core level ("chemical shift"), which in turn shifts the energy of an Auger peak involving this core level.

b) Variations of the electronic states of the valence electrons lead to changes in the peak shapes of WXV and WVV transitions.

c) Changes in the loss mechanisms alter the structure on the low-energy tail of an Auger peak.

Usually two or all three of these effects are involved and they are then difficult to separate. Chemical shifts may concern all three atomic levels which are involved in the Auger process. The occurrence of such chemical shifts of up to several eV in the energies of core levels has been frequently demonstrated by the ESCA method [24]. It is preferable to study such effects by a one-electron method, e.g. photoelectron spectroscopy. Mechanism b) also contains the very interesting effect of new electronic states produced by the formation of chemisorption bonds, but again such questions are better studied by other techniques (photoelectron spectroscopy, Auger neutralization spectroscopy etc.).

In spite of the limited resolution of most Auger spectrometers energy shifts and variations in peak shape have been observed, especially after oxidation or carbonisation of the surface. Taking Auger spectra with high instrumental resolution [44] would however be preferable in any case.

Harris [8] observed a 9 eV-shift of the energy of the O-Auger signals by comparing oxidized Al with saphire (a bulk Al_2O_3) which could not be explained by surface charging effects. Dramatic effects during the oxidation of Be are reported by Fortner and Musket [41]. One of the peaks of the Be doublet (92 and 104 eV $\triangleq 1s2s2s$- and $1s2p2p$-transitions) disappears and the other is shifted by 3 eV. Also the oxidation of Ta [42] and Mn [43] causes shifts in the Auger peaks and pronounced variations of their relative intensities.

Recently Szalkowski and Somorjai [87] made a systematic investigation of the oxidation of vanadium. Using both the chemical shifts and the oxygen-to-vanadium Auger intensity ratios the oxidation of V metal to VO and then to V_3O_5 was observed. Fig. 2.14 shows the chemical shifts of the vanadium $L_2M_{2,3}M_{2,3}$ Auger transition as a function of the oxygen to vanadium peak height ratio, h_o/h_v. The filled circles represent the known vanadium oxides that were used as reference points.

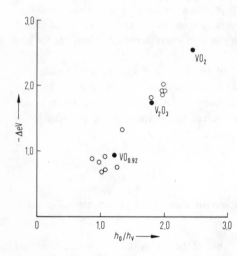

Fig. 2.14. Chemical shifts of the vanadium $L_2M_{2,3}M_{2,3}$ Auger transitions versus the oxygen to vanadium peak height ratio, h_0/h_v, for vanadium with different states of oxidation. After Szalkowski and Somorjai [87].

In special cases it is possible to obtain further information about the chemical nature of adsorbed species. For example the shape of the carbon-Auger peak on Mo(110) for adsorbed CO differs from that of elemental carbon [45].

A detailed study with the C/Ni system has been made by Coad and Rivière [46], who concluded that a carbon overlayer on Ni transforms irreversibly from a Ni_3C to a graphite layer above 400 °C. This picture is consistent with structural transformations on carbon covered Ni(110) as observed by LEED where the formation of graphite as the final stage was detected [47].

The operation of effect c) has been demonstrated by Palmberg [43] in the case of Mn and MnO. A detailed explanation of the observed effects is still lacking, since the experimental evidence is scarce and because of the rather complicated nature of the processes.

2.5. Intensity of the Auger electron emission

For a free atom the intensity of a given Auger peak is determined by the probability Q_i (ionization cross section) for the creation of a core hole in the energy level E_i and by the probability P_a for the emission of an Auger electron due to the filling of this vacancy:

$$I \propto Q_i \cdot P_a \tag{2.3}$$

2.5.1. The Auger yield

If a core is filled and an Auger electron is emitted with a probability P_a, then $P_x = 1 - P_a$ is the probability for the emission of x-rays, since this is the only competing mode for de-excitation.

The Auger yield (including Coster-Kronig transitions) is then defined by

$$Y_a = \frac{P_a}{P_a + P_x}$$

(2.4a)

and the x-ray yield by

$$Y_x = \frac{P_x}{P_a + P_x}$$

(2.4b)

For transitions of the K type (x-ray emission K_α, K_β ...., Auger emission KLL) the probability for x-ray fluorescence should be proportional to Z^4, since this process is caused by an electric dipole interaction and the electric fields created by the electron as well as by the hole are proportional to Z^2 [48].

The probability for an Auger transition can be estimated on the following basis. The electrons involved in the process are described by wave functions ψ_1 and ψ_2 before the transition takes place and by ψ_1' and ψ_2' afterwards. (If electron 2 is emitted then ψ_2' describes a free electron and ψ_1' an electron in the doubly ionized atom). The interaction between the two electrons is given by the electrostatic potential $U = \dfrac{e^2}{r_{12}}$, where r_{12} is the mutual distance between the electrons. The probability P_a then follows from

$$P_a = [\iint \psi_1'^* \psi_2'^* U \psi_1 \psi_2 \, d\tau_1 \, d\tau_2]^2$$

(2.5)

If the wave functions of electrons of inner shells are assumed to be hydrogen-like it can be shown that P_a is independent of the nuclear charge Z [49].

$$Y_a = \frac{1}{1 + \beta Z^4} \; ; \qquad Y_x = \frac{\beta Z^4}{1 + \beta Z^4}$$

(2.6)

The parameter β has to be adjusted following experimental data.

An improved semi-empirical relation for K transitions has been derived by Burhop [50]:

$$\frac{1 - Y_a}{Y_a} = (-a + bZ - cZ^3)^4$$

(2.7)

Using the numerical values: $a = 6.4 \cdot 10^{-2}$, $b = 3.4 \cdot 10^{-2}$, $c = 1.03 \cdot 10^{-6}$, the curve plotted in fig. 2.15 results. It can be seen that up to $K(Z=19)$ more than 90% of the K-emission takes place via Auger emission; the probabilities for x-ray fluorescence and Auger emission are equal for Ge ($Z=32$) and then with further increasing atomic number the Auger yield decreases rapidly. However as discussed in the second section for elements with atomic numbers above ~ 13 it is convenient in surface studies to look at L, M ... Auger transitions. Here the theoretical situation is more complicated, but experiments indicate that for L-transitions the Auger yield is larger than 90% up to atomic number 50. In general it can be concluded that the production of x-rays is negligible especially for lighter elements and energies below ~ 500 eV and becomes comparable to the Auger yield only in the energy range of about 2000 eV [2], [5]. This favourable situation is one of the major reasons why Auger spectroscopy is such a sensitive tool.

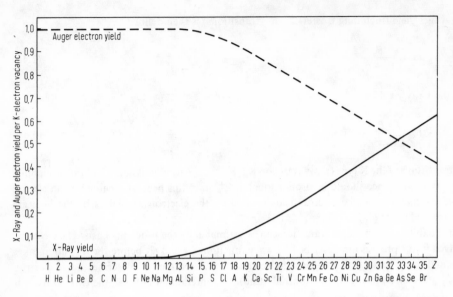

Fig. 2.15. Auger electron and x-ray yields per K-electron vacancy as a function of the atomic number. After Siegbahn et al. [24b].

2.5.2. The ionization cross section

The ionization cross section Q_i for excitation by electron impact depends strongly on the energy E_i of the bound electron as well as on the primary energy E_p of the impinging electron beam. A theoretical treatment by Worthington and Tomlin [51] leads to the relation

$$Q_i = \frac{2\pi e^2}{E_p \cdot E_i} \cdot b \ln \frac{4 E_p}{B} \tag{2.8}$$

where e is the electronic charge; $b = 0.35$ for K electrons and $b = 0.25$ for L electrons; $B \equiv (1.65 + 2.35 \exp [1 - E_p/E_i]) E_i$. To a rough approximation this relation may also be written as [21]

$$Q_i \propto \frac{f(E_p/E_i)}{E_i^2} \tag{2.9}$$

The function $f(E_p/E_i)$ contains the dependence on E_p and should be valid for all Auger transitions. It is shown in fig. 2.16 and can be seen to have a maximum at about $E_p \approx 3 E_i$ in accordance with theory [54, 55]. It is easy to explain this behaviour qualitatively: For $E_p < E_i$, the primary radiation has not enough energy for ionization and for $E_p \gg E_i$ the primary electron is too fast for appreciable interaction with the bound electron. It is therefore reasonable to use primary electrons for the ionization in the range of a few keV (usually 3 keV) in order to optimize the Auger yields for the transitions at a few hundred eV, which are those normally observed.

The second important consequence of the above relation is the rapid decrease of the ionization

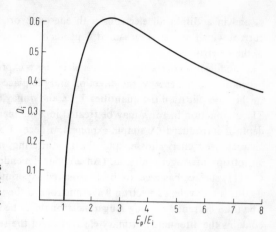

Fig. 2.16. Variation of the ionization cross section Q_i in arbitrary units of a level at energy E_i with the ratio E_p/E_i, where E_p is the energy of the primary electrons. After Bishop and Rivière [61].

cross section with increasing energy E_i of the bound electron. At the maximum of $f(E_p/E_i)$ we obtain [20]

$$Q_{i, \, max} \approx \frac{250}{E_i} \, [\text{Å}^2] \ (E_i \text{ in [eV]}).\qquad(2.10)$$

For example in order to obtain $Q_i > 10^{-20}$ cm^2 it is necessary that $E_i < 1600$ eV and this is just the energy range chosen for AES for other reasons, namely the need for a high Auger yield and a small escape depth of the Auger electrons. It is evident that AES is favoured by the coincidence of the optima of several parameters.

Experimental determinations of ionization cross sections have been made by several investigators [52], [53]. The use of atomic ionization cross sections together with (known) Auger yields should allow the surface concentrations of elements to be determined quantitatively. However, for condensed matter further complications arise making quantitative analysis difficult.

2.5.3. Auger electron emission from condensed matter

One can try to determine the intensity of the Auger current emitted from an atom in a solid ab initio. Compared with a free atom at least two additional complicating factors must be taken into account, namely the modification of the incident beam and of the Auger electrons on passing through the solid.

Following Bauer [20] the current di_a originating from atoms with concentration $c_i(z)$ in a layer of thickness dz at a depth z and emitted into the solid angle $d\Omega$ may be written as

$$di_a = c_i(z) \cdot Y_a(E_a) \cdot \int_{E_i}^{E_p} i(E_p, s_0, E, z) \cdot Q_i(E) dE \cdot a(E_a, s, z) d\Omega dz\qquad(2.11)$$

E_i is the binding energy of the atomic level which is ionized with an ionization cross section $Q_i(E)$ in order to create an Auger electron with energy E_a with the Auger yield Y_a; $i(E_p, s_0, E, z)$ is the total current capable of ionization striking the atoms in the depth z. This current contains the contribution of the primary beam with energy E_p impinging on the surface with direction

s_0 and in addition all electrons with energy E originating from secondary processes. The current which in fact is measured depends of course furthermore on the transmission factor of the spectrometer in use.

The modification of the flux of Auger electrons is expressed by an attenuation factor $a(E_a, s, z)$, depending on the energy, the direction and the place of creation of the Auger electrons. As a further complication the quantities Y_a, i, and a may depend on the chemical environment.

The attenuation factor a may be treated in a manner similar to section 2.6 where the escape depth is introduced. A simple exponential decay however does not allow the possibility of characteristic energy losses and elastic scattering, an effect which could lead to a spatial anisotropy of Auger emission. (An angular dependence of Auger electron emission from a Cu(111) surface has recently been observed experimentally [56]. This effect can probably be used to study valence electron wavefunctions in the surface region).

The total current i exciting Auger electrons may be separated into two parts: The first term contains the attenuated primary electrons and the second the scattered electrons. The effect of backscattered electrons may be very important and can lead to an enhanced Auger emission of an atom in condensed matter compared with that of the free atom. This effect is clearly demonstrated for example by the work of Tarng and Wehner [57] who studied the Auger spectra of Mo sputtered onto W. The intensity of the W-Auger signal decreases with an

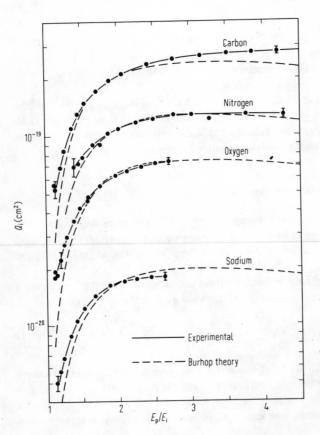

Fig. 2.17. Total K-shell ionization cross sections for C, N, O and Na adsorbed on W(100) as a function of E_p/E_i. The dashed lines were computed from the first Born approximation theory. After Gerlach and DuCharme [60 b].

exponential law with increasing overlayer thickness as predicted by the simple attenuation model. On the other hand the intensity of the Mo-120 eV-Auger electrons does not approach the value for bulk material monotonically, but has a maximum at about 7 monolayers of Mo on W which is about 20% larger than the Auger current emitted from bulk molybdenum. This surprising result is to be explained by the fact that the backscattering coefficient of W exceeds that of Mo leading to an increased current for the excitation of Auger electrons in a Mo layer on a W substrate. It has further been shown with this system that backscattering effects may be neglected for Auger electrons with higher energies.

In order to achieve a truly quantitative analysis on the basis of the Auger electron intensity it is therefore necessary to include the backscattering effects. Such attempts have been made by Gallon [58] and by Neave et al. [59] for bulk silicon. Gerlach and Du Charme [60] have determined the ionization cross sections for C, N, O and Na adsorbed on W(100) and evaluated a backscattering contribution in the case of C of about 25% for $E_p/E_i < 4$. Their results for the ionization cross sections of adsorbed atoms agree well with the data of Glupe and Mehlhorn [53] for the gaseous atoms and are drawn in fig. 2.17. For $E_p/E_i = 4$ the ionization cross sections lay between $1 \cdot 10^{-19}$ and $2 \cdot 10^{-20}$ cm^2. (In an earlier work Bishop and Rivière [61] determined the ionization cross section of the K-shell of oxygen adsorbed on Cu for $E_p/E_i = 3$ to $Q_i = 8.8 \cdot 10^{-20}$ cm^2 using the formula of Worthington and Tomlin [51].)

2.6. The detected volume

Even in the first papers using the LEED/AES technique the authors were interested to know from which depth in the solid information is obtained. This is of course a function of the electron energy and to a much smaller extent of the type of material. Since the energy of the primary electrons is usually at least three times the energy of the created Auger electrons, the mean free path of the latter ("escape depth") and not of the primary radiation is depth determining.

Weber and Peria [11] evaporated alkali metals (K, Cs) from a zeolite source onto Ge and Si single crystal surfaces up to about a monolayer. The Auger spectra were very sensitive to low surface concentrations as can be seen in fig. 2.18 for Cs/Si(100).

Similar experiments by Palmberg and Rhodin [12] for Ag on Au revealed that less than 10% of a monolayer may be detected and that at about 5 monolayers of adsorbate the contribution of the substrate becomes negligible. The simplest attempt to describe the relation between Auger emission current i_a and surface layer thickness z was using a homogeneous attenuation model. Thus

$$i_a = i_0(1 - e^{-z/z_0}) \qquad (2.12)$$

where z_0 is the escape depth of the Auger electrons characterizing their mean free path in the solid. Conversely the Auger emission current i_a' from an adsorbate covered substrate varies with the thickness z of the adsorbate layer as

$$i_a' = i_0' e^{-z/z_0} \qquad (2.13)$$

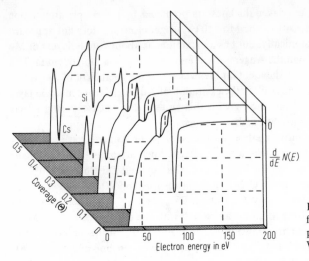

Fig. 2.18. Auger spectra for Si(100) a function of Cs coverage. The energy of primary electrons was $E_p = 300$ eV. A Weber and Peria [11].

An alternative model was proposed by Gallon [58] using a layer by layer model leading to a somewhat different relation:

$$i_n = i_\infty \cdot \left[1 - \left(1 - \frac{i_1}{i_\infty} \right)^n \right],$$

(2.14)

where i_∞ is the current emitted from bulk material, i_1 that of a single monolayer of the same material and i_n the current originating from a sheet consisting of n monolayers.

From the measured Auger peak heights in the system Ag on Au the escape depths of Auger electrons from Ag have been determined to be ~ 4 Å ($E = 72$ eV) and ~ 8 Å ($E = 362$ eV) [12]. Experiments by Tarng and Wehner [57] on the mean free path of Auger electrons in Mo and W demonstrated the validity of an exponential relation like equ. (2.12).

As can be seen from fig. 1.2 the mean free path of electrons in solid matter is not very dependent on the particular element; an estimate for the escape depth and therefore of the thickness of the analyzed surface layer may be obtained from this diagram by knowing the energy of the electrons carrying the desired information.

More exact data may be obtained by experiments with a continuous increase of the surface coverage where the Auger peak heights are calibrated against an independent absolute measure (e. g. by microgravimetry). Treating the results by Gallon's formula gives the relative contribution of the topmost layers to the total measured signal. It was determined for example that in the case of Fe evaporated onto Si about 60% of the Fe – 46 eV – Auger electrons originate from the first three atomic layers. Within the same overlayer thickness the intensity of the Si Auger electrons is reduced by about 50% [23].

The above considerations are obviously only valid if the overlayers are homogeneously distributed on the surface. This will frequently not be the case since three dimensional clusters may be formed or the adsorbate may be partially dissolved in the substrate layers near the surface. However some idea of the depth distribution may be obtained from measurements of Auger signal height with varying angle of incidence of the primary beam, as discussed in section 2.8.3.

2.7. Qualitative analysis

A surface can be analysed qualitatively simply by comparing the energies of observed Auger peaks with listed values e.g. those in fig. 2.13. In most cases the elements can be determined uniquely, even if several are present on the surface. Hydrogen and helium of course are not detected. The $dN(E)/dE$ curves contain also other characteristic peaks such as those originating from ionization losses or plasmon losses, so that not all observed peaks can be identified as Auger transitions. It is possible to separate Auger and loss peaks experimentally as has been described in section 2.2 [28].

Auger spectra have been recorded for nearly all elements including metals, semiconductors, insulators, and even organic macromolecules. Examples may be found in different places in this book. In most cases the method is only used for monitoring the chemical composition of clean or contaminated surfaces. Since nearly all LEED systems are equipped with additional facilities for measuring Auger spectra a check of the surface cleanliness by AES is the starting point of most surface investigations.

A series of earlier results was correctly interpreted only after the introduction of AES. For example a $\sqrt{19}$-superstructure observed by LEED with a cleaved Si(111) plane was formerly thought to be a property of the clean surface, but was demonstrated by AES to be caused by small amounts of nickel on the surface [62], [63].

A similar error was made with a $c2 \times 2$-structure on Ni(110) which was later shown to be caused by sulfur [64]. In the same paper Sickafus also demonstrated the possibility of using the AES method for kinetic studies. Numerous examples of the application of AES as a tool in qualitative surface analysis in connection with problems in adsorption, epitaxy, semiconductor technology, metallurgy etc. are to be found in the literature.

2.8. Quantitative analysis

The quantitative determination of surface composition presents more of a problem than qualitative analysis and there is at present no universally applicable technique for extracting absolute data from Auger spectra.

2.8.1. Determination of relative surface quantities

The relative amount of a species on the surface is fairly simple to determine because the emission current of Auger electrons is proportional to the number of excited atoms n_i. If the Auger peak in the $N(E)$ curve is Gaussian then the peak to peak height h_i of the differentiated Auger line (in the $dN(E)/dE$ curve) is proportional to n_i [65], and this quantity is frequently used as a measure of relative surface concentrations. Since the Auger current depends also on several other experimental parameters such as the incidence angle of the primary electrons, primary beam energy, modulation voltage etc., it is clear that the relative measurement is only applicable when all other parameters are held constant. A further requirement is that the arrangement of the surface atoms with respect to the substrate atoms and their chemical nature does not change with increasing coverage. For example the growth of three-dimensional nuclei or a partial incorporation into the bulk will lead to misleading results. The linearity of the peak height/coverage relation may be checked by comparison with the results of another surface sensitive technique like radioactive tracer method[35], ion counting[88], ellipsometry [89], micro balance [23], [68], proton-excited x-ray emission [65], work function measurements or flash desorption spectra.

This use of the Auger peak height as a relative measure of the surface concentration has been of great value in many investigations.

2.8.2. Absolute surface quantities

The determination of absolute surface concentrations (i.e. number of particles per cm^2) from the Auger peak heights should in principle be possible on the basis of a knowledge of all factors which contribute to the emitted current. As outlined in section 5 this problem is very complex and is not yet completely solved. Consequently the use of reference data from independent calibrations is necessary for determining absolute surface concentrations. Such independent calibration points may for example be obtained from the following techniques:

a) LEED. If from the LEED pattern the unit cell of the adsorbate structure and the positions of the adsorbed particles relative to each other can be unequivocally determined (which is not always the case) then also the surface density is known and may be compared with the Auger signal height. It is further possible to correlate this quantity with the intensity of the LEED 'extra' spots at varying coverages. An outstanding example of this kind of work is the adsorption of Xe on Pd(100) as studied by Palmberg [67].

b) Radioactive tracer. This method was used by Perdereau for the S/Ni system to determine the surface concentration of adsorbed sulfur [66].

c) Evaporated films. The thickness of evaporated metal films may be determined by micro-balance or by measuring the evaporation rate (e.g. by ion counting for the alkali metals).

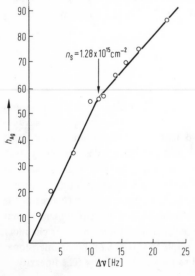

Fig. 2.19. Variation of the Auger peak-to-peak amplitude h_{Ag} for Ag on W(110) with the amount of deposited silver as recorded by the frequency change Δv of a quartz microbalance [68].

Calibrations of this type have been made with several systems. For example fig. 2.19 shows the dependence of the Auger peak height of silver deposited onto W(110) on the amount of metal evaporated as measured by a quartz microbalance [68].

d) Ellipsometry. Meyer and Vrakking [89] used an ellipsometer to determine the absolute surface quantities of several gases adsorbed on Ge and Si. The monolayer capacity was deduced from the occurrence of characteristic breaks in the relations between the optical constants; the validity of this assumption was checked by independent gas-volumetric measurements with powder surfaces.

As with all relative measurements the calibrations are only valid for one set of experimental parameters and are not readily transferable from one experiment to another.

2.8.3. Alloys

The surface composition of alloys is a very important parameter for several technological processes, particularly since it may deviate considerably from that of the bulk, for example after special treatment of the surface. Such surface compositions may be determined from Auger spectra in such manners as proposed by Ertl and Küppers [69] and Quinto et al. [70] for binary Cu/Ni alloys. This is possible if it is assumed that the Auger intensities for the pure components may be extrapolated linearly to give the concentrations in the alloys. This can best be done by choosing Auger electron energies which occur only for one element and are well separated from other transitions so that no interference takes place. An example is shown in fig. 2.20, where the Auger spectra of Ag, Pd and an Ag/Pd alloy are plotted [71]. It can be

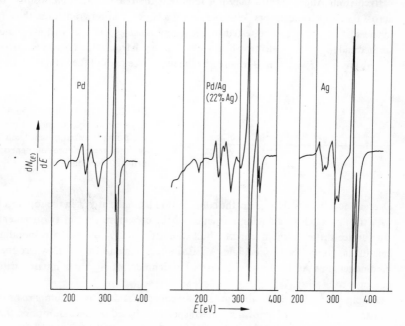

Fig. 2.20. Auger spectra from Ag, Pd and an Ag/Pd alloy containing 22% Ag. After [71].

seen that the peak at 356 eV originates only from Ag, that at 331 eV only from Pd, whereas some other transitions are superpositions of contributions from both components. The concentration x of Ag in the surface region can be determined by comparing the height of the 356 eV peak $^0h^{Ag}$ from pure Ag with that $^xh^{Ag}$ for the alloy in question using the relation

$$^xh^{Ag} = x \cdot {^0h^{Ag}}, \tag{2.15}$$

provided all experimental parameters (primary energy and current, sample geometry etc.) are held constant. A check can be made by using the Auger peak heights h^{Pd} for Pd which have to obey the relation

$$^xh^{Pd} = (1-x) \cdot {^0h^{Pd}}. \tag{2.16}$$

In this example the results were completely consistent and the surface compositions so determined agreed within the limits of error ($\sim 2\%$) with the bulk composition. However this latter result is by no means of general validity. In particular for Cu/Ni alloys it has been demonstrated that the surface treatment may alter the surface composition considerably. For example selective sputtering of Cu leading to an enrichment of Ni at the surface has been reported [70], but careful annealing at $T = 500$ °C of bulk samples restored the homogeneity [69], [72]. Lack of miscibility throughout the temperature range and thermodynamically favoured surface enrichment (Gibbs adsorption) can also lead to variations between surface and bulk composition.

Since the escape depth of the Auger electrons is of the order of a few atomic layers compositions derived from Auger spectra may not represent the real surface composition but an average across the first atomic layers. Whether or not a concentration gradient in the direction normal to the surface exists can be tested by recording Auger spectra at various angles between the surface and the primary electron beam [69]. The depth from which Auger electrons are excited is simply given by $z_a = l_p \cdot \sin \vartheta$, where l_p is the mean free path of the primary electrons (fig. 2.21).

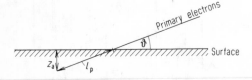

Fig. 2.21. The depth Z_a from which Auger electrons are excited depends on the angle ϑ between the primary electron beam and the surface and on the mean depth of penetration l_p of the primary electrons.

Decreasing ϑ lowers also z_a, i.e. the contribution from the atoms in the topmost layer increases with the approach to glancing incidence. If the concentrations of two elements c_1 and c_2 at the surface differ from that in the bulk the ratio $h_1 : h_2$ of the corresponding Auger peak heights has to vary with the angle. An example is shown in fig. 2.22 where the results for an inhomogeneous Ag/Pd alloy with surface enrichment of Ag are compared with those from a homogeneous alloy [71].

The method of determining absolute surface concentrations from a comparison of experimental Auger spectra with well-known standard spectra is based on the assumption that the chemical environment does not influence the Auger currents emitted from the individual species. This procedure has been shown to be justified for certain binary alloys composed of elements

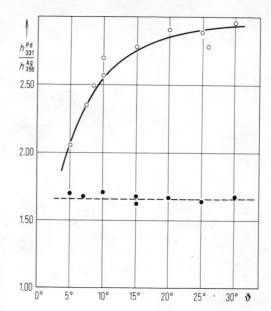

Fig. 2.22. Variation of the ratio of Auger peak heights h^{Pd}/h^{Ag} with the angle ϑ of the primary beam for a homogeneous Ag/Pd alloy (full circles) and a sample with enrichment of Ag at the surface (open circles). After [71].

which are neighbours in the periodic table. In order to generalize this method many standards with known surface compositions are needed [43] which becomes a formidable task for multi-component systems. In such cases the determination of variations of relative surface quantities alone may be very informative. This was shown in a series of experiments on the segregation of impurities at grain boundaries causing embrittlement. The first studies of this type were made by Marcus and Palmberg [73] who fractured steel samples in situ under UHV conditions. The Auger spectra of the embrittled samples showed a large antimony peak and also enlarged concentrations of Ni, Cr, C and P. It was concluded that the segregation of Sb at the grain boundaries (the overall concentration of this element was about 0.03%) is mainly responsible for the altered mechanical properties. Similar studies have been made with other steels [74, 76] and with copper [75]. In the latter case a very pronounced enrichment of Bi at the grain boundaries with a thickness of several atomic layers was observed.

2.8.4. Depth profiling

A very interesting application of AES has been described by Palmberg [77] in which a surface is sputtered continuously by noble gas ions (ion etching) and simultaneously Auger spectra are recorded thus providing a layer by layer analysis. Depth profiles are obtained by sputtering the sample with an Ar^+ ion beam of a few square millimeters cross section while the Auger spectra are recorded from a small area of less than 0.01 mm² in the centre of the sputtered region in order to insure a uniform depth across the analyzed spot. The composition profile of a silicon wafer which was covered with phosphorus and subsequently oxidized is shown in fig. 2.23. The formation of a silicon oxide layer about 2000 Å thick on top of the pure Si can be seen. The depth resolution was a few hundred Ångstroms in this example. Since the removal of material from surfaces by noble gas ion sputtering is a well developed technique

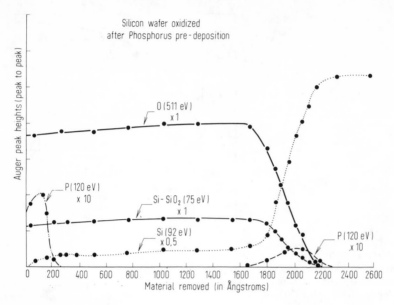

Fig. 2.23. Composition profiles for Si, P, O and C on oxidized silicon after predeposition of P. After Palmberg [43].

[78] this method offers new possibilities for studying interface reactions and problems of semiconductor technology etc. But since sputtering rates may differ considerably care has to be taken when estimating the depth of the profiles. Furthermore it is possible (particularly with uneven starting surfaces) that the erosion is far from even so that in any given spectrum a contribution from the remnants of upper layers is present. On the other hand this technique may also be used to determine sputtering rates if the thickness of the overlayer is known [79], [80].

2.8.5. Kinetic studies

Monitoring the surface concentration of a species as a function of time enables the study of the kinetics of surface processes, provided that interference of the primary electron beam with the adsorbate can be neglected. For this reason it is desirable to perform such investigations with a cylindrical mirror analyzer and low primary beam currents. In all experiments the Auger peak height is used as a relative measure for the surface concentrations.

Florio and Robertson [81] investigated the adsorption of chlorine on Si(111) using a retarding field analyzer. Electron beam stimulated desorption could not be excluded and a cross section of about $2 \cdot 10^{-19}$ cm^2 for chlorine was determined for this desorption process.

A very detailed study on the kinetics of oxidation of sulfur on a Cu(110) surface has been performed by Bonzel [82]. At the beginning of the reaction the surface was covered with sulfur by exposing the sample to H_2S. Then with different oxygen partial pressures and different temperatures the variation of the S-Auger peak with time was followed by means of a cylindrical mirror analyzer. The overall reaction $S_{ad} + O_{2, gas} \rightarrow SO_{2, gas}$ was postulated to proceed in the following three steps:

$$O_{2,\,g} + \;*\; \to 2O_{ad}$$

$$O_{ad} + S_{ad} \to SO_{ad}$$

$$SO_{ad} + O_{ad} \to SO_{2,\,g}\,,$$

where * denotes a free adsorption site. The removal of sulfur from the surface is accelerated by increase of the oxygen pressure. The proposed reaction scheme could be supported by a detailed analysis of the experimental results. Bonzel [82] showed that at lower temperatures the rate of SO_2 formation is proportional to $[O_{ad}]$, at higher temperatures proportional to $[O_{ad}]^2$. From the evaluated activation energies it was concluded that in all cases processes of surface diffusion are rate-determining. The time-constants with these examples were always of the order of several minutes. The CMA enables the kinetics of much faster surface processes with time constants down to a few milliseconds to be followed. A fast laser induced desorption technique [83] would provide a means of initiating such fast surface reactions.

2.9. Deconvolution technique and band structure

The energy distributions measured in usual electron beam excited AES do not represent the true Auger electron energy distributions but are influenced by effects of limited instrumental resolution and by loss processes within the solid. Similar to other spectroscopic methods a decon-

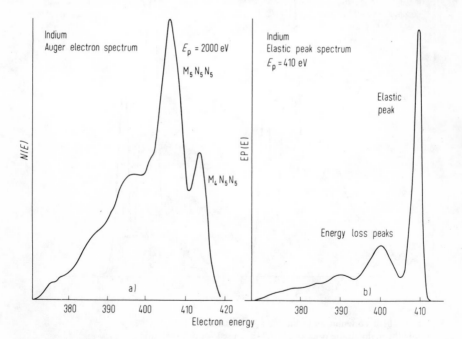

Fig. 2.24. a) Observed Auger energy distribution $N(E)$ of the In–$M_{4,5}N_5N_5$ transitions. b) Spectrum of the elastic peak at 410 eV primary energy. After Mularie and Peria [84].

volution technique has been proposed by Mularie and Peria [84] in order to extract the true energy distribution from measured Auger spectra. It must be remembered that the Auger peaks may contain additional structure from characteristic energy losses, as studied in some detail by Mularie and Rusch [85], and are further broadened by electron-electron and electron-phonon interactions. These features are then superimposed by a (larger) instrumental broadening.

If $N(E)$ is the true electron distribution and $N^*(E)$ the measured distribution then both are correlated by

$$N^*(E) = \int_{-\infty}^{+\infty} N(E') \cdot B(E-E') \, dE' \tag{2.17}$$

where the convolution function $B(E)$ contains all instrumental and inelastic effects.

A theoretical determination of $B(E)$ is at present not possible. But this difficulty may be circumvented by measuring the energy distribution around the elastically scattered peak. Then it is assumed that an Auger peak exhibits the same structure, provided that the previously measured elastic peak was at the same energy as this Auger peak. This is a somewhat weak point, for an Auger electron and a primary electron will not be equivalent since the final state of the atom which emitted the Auger electron is an excited one which may involve the creation of plasmons. Some results obtained with this technique are shown in fig. 2.24. The experimental energy distribution $N(E)$ from sputtered In as evaluated from the $dN(E)/dE$ curve measured

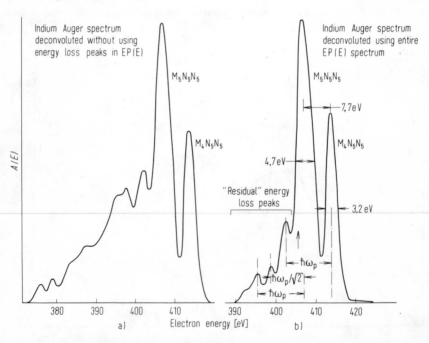

Fig. 2.25. The deconvolution technique applied to fig. 2.24.
a) If only the main elastic peak of fig. 2.24b is used for deconvolution.
b) The In–$M_{4,5}N_5N_5$ Auger spectrum, deconvoluted using the entire spectrum of fig. 2.24b.
After Mularie and Peria [84].

by an RFA by numerical integration is drawn in fig. 2.24a. The dominating features are the $M_4 N_5 N_5$ and $M_5 N_5 N_5$ transitions in the energy range of 410 eV. The energy distribution around elastically reflected electrons with primary energies of 410 eV are shown in fig. 2.24b. If only the shape of the elastic peak is used for $B(E)$ then the deconvoluted distribution of fig. 2.25a results. If the characteristic losses of the primaries are included one arrives at the spectrum of fig. 2.25b. The $M_4 N_5 N_5$ and $M_5 N_5 N_5$ transitions are now well separated and the "residual" energy loss peaks emerge from the background. Similar results have been obtained with Cd and with oxygen either adsorbed on Mo or in the form of TiO_2. The latter results agree very well with direct determinations of the energy distribution by means of ESCA [24]. If the Auger electrons are not emitted from deep lying states but from the valence band, then the above discussed treatment may be used to determine the band structure, i.e. the density of states in the band. Amelio [86] describes a method, by which the Auger signal from LVV transitions of Si may be deconvoluted stepwise thus yielding the transition probabilities for different $LV(E)V(E')$ processes as well as the density of states in the valence band. Results obtained from theoretical calculations and x-ray emission measurements are compared with those obtained with the above method in fig. 2.26.

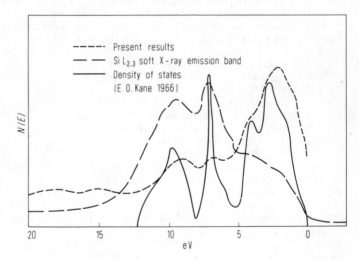

Fig. 2.26. Density of states of the valence band of Si. Comparison of theoretical and x-ray emission data with results from AES. After Amelio [86].

2.10. References

[1] P. Auger, J. Phys. Radium *6*, 205 (1925).
[2] E.H.S. Burhop: The Auger effect and other radiationless transitions. Cambridge Univ. Press., London 1952.
[3] W.N. Asaad, Nucl. Phys. *5/65*, 494 (1966).
[4] O. Hörnfeldt, A. Fahlmann and C. Nordling, Arkiv Fysik *23*, 155 (1962).
[5] I. Bergström and C. Nordling, in: Alpha-, Beta- and Gamma-Ray Spectroscopy. Vol. 2 (K. Siegbahn, ed.), North Holland, Amsterdam 1965.
[6] J.J. Lander, Phys. Rev. *91*, 1382 (1953).
[7] L.N. Tharp and E.J. Scheibner, J. Appl. Phys. *38*, 4355 (1967).
[8] L.A. Harris, J. Appl. Phys. *39*, 1419 (1968).
[9] L.A. Harris, J. Appl. Phys. *39*, 1428 (1968).

[10] L. B. Leder and J. A. Simpson, Rev. Sci. Instr. *29*, 571 (1958).
[11] R. E. Weber and W. T. Peria, J. Appl. Phys. *38*, 4355 (1967).
[12] P. W. Palmberg and T. N. Rhodin, J. Appl. Phys. *39*, 2425 (1968).
[13] P. W. Palmberg, Appl. Phys. Lett. *13*, 183 (1968).
[14] P. W. Palmberg, G. K. Bohn and J. C. Tracy, Appl. Phys. Lett. *15*, 254 (1969).
[15] T. W. Haas, G. J. Grant, A. G. Jackson and M. P. Hooker: A. bibliography of LEED and AES. in: Progress in Surface Science. Vol. 1 (2), Pergamon Press, Oxford 1971.
[16] D. T. Hawkins: Bibliography of Auger Electron Spectroscopy. Bell Telephone Library. March 1972, No. 201.
[17] N. J. Taylor, in: Techniques of Metals Research (Ed. R. F. Bunshah), Vol. 7, Interscience Publ., NewYork 1971.
[18] C. C. Chang, Surface Sci. *25*, 53 1971.
[19] E. N. Sickafus and H. P. Bonzel, in: Recent Progress in Surface Science (J. F. Danielli, K. G. A. Pankhurst and A. C. Riddiford, eds.). Vol. 4, Academic Press, New York 1971.
[20] E. Bauer, Vacuum *22*, 539 (1973).
[21] J. C. Tracy, in Electron Emission Spectroscopy, (W. Dekeyser et al., eds.), Reidel Publ. Comp., Dordrecht 1973, p. 295.
[22] P. W. Palmberg, in: Electron Spectroscopy (Ed. D. A. Shirley), North Holland, Amsterdam 1972, p. 835.
[23] J. W. T. Ridgeway and D. Haneman, Surface Sci. *24*, 451 (1971).
[24] a) K. Siegbahn et al.: ESCA. Atomic, Molecular and Solid State Structure Studied by means of Electron Spectroscopy. Almquist & Wiksells, Uppsala 1967.
b) K. Siegbahn et al.: ESCA Applied to Free Molecules. North Holland, Amsterdam 1969.
[25] D. L. Smith and D. A. Huchital, J. Appl. Phys. *43*, 2624 (1972).
[26] R. G. Musket and W. Bauer, Appl. Phys. Lett. *20*, 455 (1972).
[27] D. K. Skinner and R. R. Willis, Rev. Sci. Instr. *43*, 731 (1972).
[28] R. L. Gerlach, J. E. Houston and R. L. Park, Appl. Phys. Lett. *16*, 179 (1970).
[29] E. N. Sickafus, Rev. Sci. Instr. *42*, 933 (1971).
[30] L. Fiermans and J. Vennik, Surface Sci *38*, 237 (1973).
[31] E. N. Sickafus and A. D. Colvin, Rev. Sci. Instr. *41*, 1345 (1970).
[32] J. R. Arthur, J. Vac. Sci. Techn. *10*, 136 (1973).
[33] H. P. Bonzel, J. Chem. Phys. *59*, 1641 (1973).
[34] Varian Corp., Palo Alto (Calif.).
[35] N. C. Mac Donald and J. R. Waldtrop, Appl. Phys. Lett. *19*, 315 (1971).
[36] D. Coster and R. L. Kronig, Physics *2*, 13 (1935).
[37] W. N. Asaad and E. H. S. Burhop, Proc. Phys. Soc. *71*, 369 (1958).
[38] I. Bergström and R. D. Hill, Arkiv Fysik *8*, 21 (1954).
[39] Y. Strausser and J. J. Uebbing: Varian Chart of Auger Electron Energies. Varian Corp., Palo Alto (USA) 1970.
[40] H. E. Bishop and J. C. Rivière, Surface Sci. *17*, 462 (1969).
[41] R. J. Fortner and R. G. Musket, Surface Sci *28*, 504 (1971).
[42] T. W. Haas and J. T. Grant, Phys. Lett. *30 A*, 272 (1969).
[43] P. W. Palmberg in: Electron Spectroscopy. (D. A. Shirley ed), North Holland, Amsterdam 1972.
[44] C. Nordling, E. Sokolowski and R. Siegbahn, Arkiv Fysik *13*, 483 (1958).
[45] T. W. Haas and J. T. Grant, Appl. Phys. Lett. *16*, 172 (1970).
[46] J. P. Coad and J. C. Rivière, Surface Sci. *25*, 609 (1971).
[47] G. Ertl in: Molecular Processes on Solid Surfaces. (Drauglis, Gretz, Jaffe eds.). McGraw Hill, New York 1969
[48] G. Wenzel, Z. Physik *43*, 524 (1927).
[49] L. D. Landau and E. M. Lifshitz: Quantenmechanik. Akademie-Verlag, Leipzig 1965.
[50] E. H. S. Burhop, J. Phys. Radium *16*, 625 (1955).
[51] C. R. Worthington and S. G. Tomlin, Proc. Phys. Soc. *A 69*, 401 (1956).
[52] W. Hink and H. Paschke, Z. Phys. *244*, 140 (1971).
[53] G. Glupe and W. Mehlhorn, Phys. Lett. *25 A*, 244 (1967).
[54] H. S. W. Massey and E. H. S. Burhop: Electronic and Ionic Impact Phenomena. Oxford Clarendon Press, New York 1952.

[55] H. W. Darwin, Z. Phys. *164*, 513 (1961).
[56] B. W. Holland, L. McDonnell and D. P. Woodruff, Solid State Comm. *11*, 991 (1971).
[57] M. L. Tarng and G. K. Wehner, Proc. Physical Electronic Conf., Albuquerque, New Mex., 1972.
[58] T. E. Gallon, Surface Sci *17*, 486 (1969).
[59] J. H. Neave, C. T. Foxon and B. A. Joyce, Surface Sci. *29*, 411 (1972).
[60] R. L. Gerlach and A. R. Du Charme, Surface Sci. *29*, 317 (1972); *32*, 329 (1972).
[61] H. E. Bishop and J. C. Rivière, J. Appl. Phys. *40*, 1740 (1969).
[62] J. M. Charig and D. K. Skinner, Surface Sci. *19*, 283 (1970).
[63] N. J. Taylor, Surface Sci. *15*, 169 (1969).
[64] E. N. Sickafus, Surface Sci. *19*, 181 (1970).
[65] P. B. Needham, T. J. Driscoll and N. G. Rao, Appl. Phys. Lett. *21*, 502 (1972).
[66] M. Perdereau, Surface Sci. *24*, 239 (1972).
[67] P. W. Palmberg, Surface Sci. *25*, 598 (1971).
[68] E. Bauer and H. Poppa, Thin Solid Films, *12*, 167 (1972).
[69] G. Ertl and J. Küppers, Surface Sci. *24*, 104 (1971).
[70] D. T. Quinto, V. S. Sundaram and W. D. Robertson, Surface Sci. *28*, 504 (1971).
[71] K. Christmann and G. Ertl, Surface Sci. *33*, 254 (1972).
[72] G. Ertl and J. Küppers, J. Vac. Sci. Techn. *9*, 829 (1972).
[73] H. L. Marcus and P. W. Palmberg, Trans. Met. Soc. AIME *245*, 1699 (1966).
[74] D. F. Stein, R. E. Weber and P. W. Palmberg, J. Met. (USA) *23*, 39 (1971).
[75] A. Joshin and D. F. Stein, J. Inst. of Metals *99*, 2449 (1971).
[76] G. J. Dooley, J. Vac. Sci. Techn. *9*, 145 (1972).
[77] P. W. Palmberg, J. Vac. Sci. Techn. *9*, 160 (1972).
[78] M. L. Tarng and G. K. Wehner, J. Appl. Phys. *43*, 2268 (1972).
[79] T. Smith, Surface Sci. *27*, 45 (1971).
[80] R. N. Yasko and L. J. Fried, Rev. Sci. Instr. *43*, 335 (1972).
[81] I. V. Florio and W. D. Robertson, Surface Sci. *18*, 398 (1969).
[82] H. P. Bonzel, Surface Sci. *27*, 387 (1971).
[83] G. Ertl and M. Neumann, Z. Naturforschung *27 a*, 1607 (1972).
[84] W. M. Mularie and W. T. Peria, Surface Sci. *26*, 125 (1971).
[85] W. M. Mularie and T. W. Rusch, Surface Sci. *19*, 469 (1970).
[86] G. F. Amelio, Surface Sci. *22*, 301 (1970).
[87] F. J. Szalkowski and G. A. Somorjai, J. Chem. Phys. *56*, 6097 (1972).
[88] S. Thomas and T. W. Haas, J. Vac. Sci. Techn. *9*, 840 (1970).
[89] F. Meyer and J. J. Vrakking, Surface Sci. *33*, 271 (1972).
[90] T. E. Gallon and J. A. D. Matthew, Rev. Phys. Technol. *3*, 31 (1972).
[91] P. F. Kane and G. B. Larrabee, Ann. Rev. Mater. Sci *2*, 33 (1972).
[92] L. G. Parratt, Rev. Sci. Instr. *30*, 297 (1959).

3. Electron Energy Loss Spectroscopy (EELS)

3.1. General remarks

Besides the elastically scattered electrons, which form the basis of the LEED method, and the Auger electrons, as described in the preceding chapter, electrons which have undergone characteristic energy losses in the surface region may be used for surface characterization. Loss peaks are created by inelastic scattering of primary electrons with energy E_p, whereby the energy E_1 is lost, leading to secondary electrons with energy $E_p - E_1$.

Electrons which have suffered energy losses by inelastic collisions may be distinguished from the other features in the energy distribution curve by using the energy E_p of the primary electrons as a reference. Peaks due to energy losses have a constant energy difference with respect to E_p, so that a variation ΔE of E_p also shifts the loss peaks by ΔE. On the other hand Auger electrons and true secondaries have fixed energies, and only the shapes of the various peaks may change on variation of E_p. It is evident that characteristic energy losses are not only undergone by primary electrons with energy E_p but also by any secondary electron. This is true for Auger electrons, giving rise to satellites in the vicinity of Auger peaks, which are however frequently too small to be detected.

The characteristic losses may be divided into the following four categories:

a) Excitation of core electrons (Ionization spectroscopy). Transitions of this type are observed if the primary electron interacts with an electron which is bound in an inner level of a surface atom and transfers an amount of energy which is sufficient to ionize the atom by exciting the core electron to an unfilled state above the Fermi level. The magnitudes of the energy losses are determined by the binding energies of the core level electrons and are therefore in the same range as the energies of the Auger electrons.

b) One-electron excitations of valence electrons. An electron in the valence band of a solid may be excited to a higher (unfilled) level of the same band (intraband transition) or into another energy band (interband transition). A further possibility arises if by the chemisorption of foreign particles additional electronic states at the surface appear and electronic transitions in the adsorbates occur. The energy losses of the primary electrons associated with such processes are typically of the order of 3–20 eV. The excited electrons have a certain probability of leaving the solid to appear in the "true secondary" electron peak at the low energy side of the backscattered energy distribution curve.

c) Collective excitations of valence electrons (Plasmon losses). These are collective oscillations of conduction electrons whose energy of excitation lies in the range between 5 and 60 eV.

d) Excitations of surface vibrations (Quasielastic electrons). The phonon-assisted inelastic scattering of electrons leads to very small energy losses in the range of less than a few tenths of an eV, so that these electrons are not separated from the elastic peak in normal spectrometers. Much more refined experimental techniques are necessary in order to determine the energies of these excitations. This subject is treated further in chapter 10.

3.2. Experimental

Characteristic energy losses may (with the exception of "quasi"-elastic electrons) be studied with the same experimental arrangement as that used for Auger electron spectroscopy. In fact for ionization spectroscopy the primary energy is similar to that used for AES, and both types of peaks are contained in the measured $N(E)$ or $dN(E)/dE$ curves. Variation of the electronic modulation technique [1] as described in section 2.2 may help to separate both features. For valence electron excitations lower primary energies of some 20–500 eV are favourable. The primary electrons are emitted from an electron gun as a focused beam with sufficient intensity and low energy spread since in contrast to AES the energy spread of the primary beam influences the width of the measured loss peaks. The most convenient source is an electron gun such as is used for LEED. The LEED grid optics may be used as retarding field analyzer, as well as the cylindrical mirror analyzer.

In more refined investigations (ILEED, see section 3.5) not the total integrated backscattered current is measured, but rather information on the angular distribution (as well as the energy distribution) is required. This is obtained using a Faraday cup similar to that used for LEED intensity measurements. A sophisticated device [2] with high angular and energy resolution is shown schematically in fig. 3.1. The underlying principle is the same as that of the original

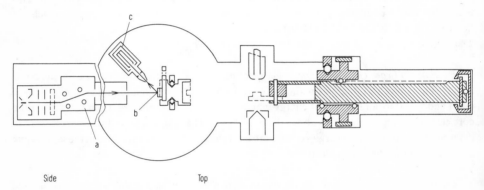

Side   Top

Fig. 3.1. Refined electron diffractometer as described by Porteus [2].
a: Electron gun with deflection plates,   b: sample,   c: collector.

apparatus used by Davisson and Germer [3] for the detection of the wave nature of slow electrons; the collector and the sample can be rotated around different axes. Another more simple device uses a LEED-RFA together with an optical spot photometer [4].

Detection of the signals is either achieved by employing the usual differentiation techniques as for AES or by direct recording of energy distribution curves.

3.3. Ionization spectroscopy

The overall mechanism of ionization spectroscopy (IS) of a metal is shown in fig. 3.2. The incident electron with energy E_1 transfers energy to an electron in a level $-E_i$ thereby exciting it to an empty state E_f above the Fermi level. The loss energy is given by

$$E_L = E_1 - E_2 = E_i + E_f \tag{3.1}$$

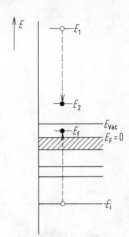

Fig. 3.2. Mechanism of ionization spectroscopy. The origin of the energy scale is taken at the Fermi level E_F.

In contrast to AES where three electronic states are involved the mechanism of ionization spectroscopy is determined by two electronic transitions. The ejected core electron may be excited to any available state above the Fermi level; the minimum loss energy for excitation of a certain core state E_i is attained for excitation just to the Fermi level. A "tail" or fine structure will be observed at higher loss energies. These features are associated with the density of unfilled states above the Fermi level E_F whereas AES probes the filled states below E_F. Such fine structure may help to characterize the chemical states of the elements on the surface.

The exciting electron can be either an electron from the primary beam with energy E_p which gives rise to the characteristic energy loss peaks [5] or an Auger electron which then causes satellites on the Auger peaks [6]. The energy distribution curves usually consist of a superposition of Auger peaks, plasmon excitations and ionization losses. Recording spectra with two different primary energies or (more convenient) using the electronic modulation technique as proposed by Gerlach et al. [1] allows a separation of the characteristic losses.

A loss spectrum obtained by means of this technique is shown in fig. 3.3 together with the corresponding Auger electron spectrum [7]. The sample was tungsten contaminated with C and O. Fig. 3.3c is the second derivative $d^2 N/d E^2$ of the IS, which exhibits more details of the fine structure [8].

The peaks correspond very well with the binding energies of the different electronic states, as determined by photoemission and indicated by arrows. The peaks are sharper than the Auger lines, their widths are mainly determined by the instrumental resolution.

This example demonstrates that IS can be used for an elemental analysis of the surface. The observed spectra are less complex than Auger spectra since fewer electronic states are involved, but the signals are much weaker than those obtained with AES and the method has found little application in surface analysis. IS spectra may be used for chemical identification in cases where Auger peaks overlap on the energy scale or where the observed Auger peaks reveal no unique evidence for the existence of an element.

The sensitivity of IS peaks to the chemical state of the analysed element is demonstrated by the

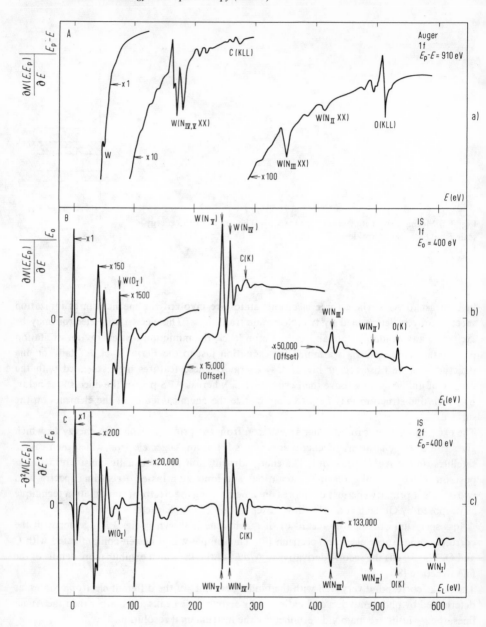

Fig. 3.3. Auger and ionization spectra for a tungsten surface contaminated by oxygen and carbon
a) Auger spectrum $dN(E)/dE$
b) Ionization spectrum (first derivative)
c) Second derivative of the ionization spectrum.
    After Gerlach [7].

spectra of fig. 3.4 [7]. Fig. 3.4a shows the spectrum of a W(100) surface heavily contaminated by carbon originating from the decomposition of hydrocarbons. After heating the crystal to about 1 200 °C the shape of the C peak had altered (fig. 3.4 b). Although no interpretation for this observed change of the signal shape can yet be given, it clearly demonstrates the influence of chemical effects.

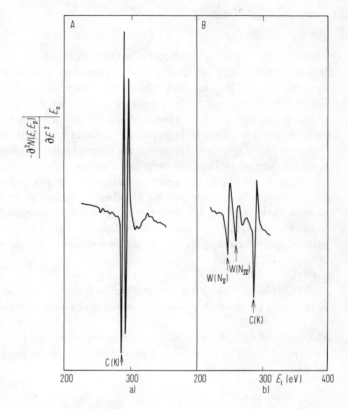

Fig. 3.4. Ionization spectra (second derivatives) of two chemical states of carbon on W(100).
a) Doublet peak structure after bakeout of the vacuum system.
b) Upon heating the sample to 1200 °C the C(K)-peak became singlet.
After Gerlach [7].

A third application of ionization spectroscopy mainly of interest to the physicist is the determination of differential ionization cross sections of surface atoms by measuring the ionization peak current which is scattered into a narrow angle $d\Omega$ [9].

3.4. Plasmon losses

As outlined in section 3.1 energy losses involving valence electrons include plasmon excitations. Plasmons are collective excitations of the free (or nearly free) electron gas, as that is the conduction electrons in metals. A plasmon (or a plasmon oscillation) may be described as an oscillation of the conduction electrons with respect to the positive ion cores of the crystal lattice with frequency ω_p.

If the conduction electrons are displaced collectively by a distance Δx against the ion cores a restoring force proportional to Δx results due to Coulomb interaction. In analogy to the mechanical system of a spring the solution of the equation of motion represents a harmonic oscillation.

The theory of plasmons has been developed by Bohm and Pines [10]. The frequency ω_p of the undamped plasma oscillation of a free electron gas is given by

$$\omega_p = \frac{4\pi n_e e^2}{m} \qquad (3.2)$$

where n_e is the density of the electrons, m their effective mass, and e the electron charge.

It is evident that such oscillations may be excited by creating an electromagnetic a.c. field by illumination of the solid with light of the appropriate wavelength. If the frequency of the light equals the frequency of the plasma oscillation energy is absorbed by the solid so that the dielectric constant $\varepsilon(\omega)$ of the medium exhibits characteristic variations in this range. It is therefore possible to determine plasmon frequencies from measurements of complex dielectric constants. The subject is rather outside the range of interest of this work and a review of such results by Raether [11] is recommended for further information.

The exact theory of plasmon excitations is rather complex and is based on a description of fluctuations in the electron gas, but the classical picture as outlined above assuming pure electrostatic interactions yields the correct result for the plasmon frequency. It has to be born in mind however, that a plasmon must be considered as an elementary excitation, i.e. a quasi-particle, characterized by its energy and momentum [12]. This leads to the classification of two types of plasmons:

a) A bulk plasmon which has a momentum with a component normal to the surface.
b) A surface plasmon which has no component of momentum normal to the surface.

The theory of surface plasmons was established by Ritchie [13] who demonstrated that the frequency of a surface plasmon ω_s depends on the dielectric constant ε' of the medium outside the solid:

$$\omega_s = \frac{\omega_p}{\sqrt{1+\varepsilon'}} \qquad (3.3)$$

where ω_p is the frequency of the bulk plasmon. If the surface is not covered by an adsorbed layer, then $\varepsilon' = 1$, yielding:

$$\omega_s = \frac{1}{\sqrt{2}} \cdot \omega_p \qquad (3.4)$$

If for example an oxide layer exists on the surface with $\varepsilon' > 1$ the frequency of the surface plasmon is lowered [14].

The difference between bulk and surface plasmons becomes more evident from the model in which the oscillating electrons are described by charge density functions $\varrho(r)$. With the surface plasmon $\varrho(r)$ decays very strongly in the direction normal to the surface, restricting the plasmon oscillations to the region near the surface, in contrast to the bulk plasmon.

Equations (3.2)–(3.4) for ω_p and ω_s are strictly valid only for a free electron gas, which is approximated only by a few metals like Al, Be or Mg. If the electrons are more tightly bound and if single-electron transitions between different states within the bands become possible then single-electron processes (inter- and intraband transitions) will compete with plasmon excitations for energy absorption [12]. A detailed investigation [15] of these effects revealed that the plasma frequencies ω_p are shifted, the magnitude of these shifts depending on the positions of the plasmon frequencies on the energy scale relative to the energies of the single-electron transitions.

Plasmons may also be created by primary electrons which penetrate into a solid creating local Coulombic fields. This is described by the dielectric theory, in which the polarization caused by the primary electron is given by the dielectric function $\varepsilon(\boldsymbol{k}, \omega)$, where $\boldsymbol{k}$ is the wave vector of the excitation and $\hbar\omega$ its energy. In 1936 Rudberg [16] detected characteristic energy losses

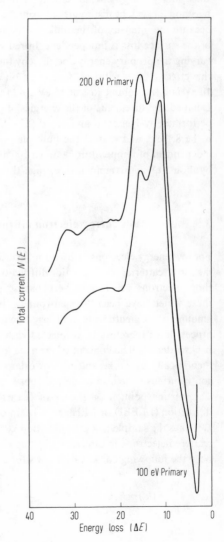

Fig. 3.5. Energy-loss spectrum for aluminium at 100 and 200 eV primary energies showing the variation of the intensity of the surface loss at 10 eV with the free mean depth of penetration of the primary electrons. After Simmons and Scheibner [28].

by reflection of slow electrons from solid surfaces before the theory had been developed. Ferrell [17] demonstrated why electrons passing through thin films may excite plasmons and the same arguments are also true for electrons reflected from a surface.

Energies for plasmon excitations are in the range of 5–60 eV. The number of transitions varies with the element. Experimental results have been reviewed by Pines [18] and by Raether [11], but most have been obtained using badly defined surfaces.

The dielectric theory may also be used to determine the optical constants of solids from energy losses of fast electrons [19]. Valence electron excitations with interesting features for surface physics are best studied with low primary energies (< 100 eV). In this case diffraction effects have to be taken into consideration which introduces further complexity.

Some plasmon peaks may be identified in fig. 3.3. These features become more pronounced if the primary energy E_p is lowered to values < 100 eV. In transmission experiments the surface-plasmon loss can be distinguished from the volume-loss by varying the thickness of the sample, because the intensity of the bulk loss increases with the film thickness whereas the intensity of the surface loss is independent. In reflection studies this criterion can be approximated by varying the primary energy and thereby the penetration depth of the electrons. Fig. 3.5 shows this effect clearly with measurements on clean Al from which the surface plasmon loss at 10 eV and the volume loss at 15 eV could be identified [28].

Detailed measurements of the energies of plasmon losses at clean Ni(111) surfaces have been performed by Heimann and Hölzl [27]. The energy of the surface plasmon was determined to be 8.1 eV, and that of the bulk plasmon 19.1 eV. The energy of the surface plasmon is independent of temperature between 100 and 700 °C, whereas the bulk plasmon loss is shifted by about 1 eV in this temperature range, this shift being nearly step-like at the Curie temperature.

3.5. Inelastic low energy electron diffraction (ILEED)

For low energy electrons as used for surface studies there is an important connection between inelastic scattering and (elastic) diffraction. This becomes obvious from the fact that only those electrons which have been backscattered may reach the collector or in other words those which have been "turned-round". In inelastic scattering processes usually only small amounts of momentum are transferred, so that forward-scattering is predominant. A diffraction process must therefore be connected with the inelastic scattering process in order to achieve backscattering. The concept of inelastic low energy diffraction (ILEED) has been developed theoretically by Duke and his coworkers [20–23] only recently, although some important contributions were also made previously by Weber and Webb [36].

The starting point for a theory of ILEED is similar to that of elastic low energy electron diffraction (ELEED or LEED) : A beam of primary electrons with energy $E(k)$ and intensity I_0 strikes the sample from the direction s_0. The problem is to calculate the intensity I of back-scattered electrons in direction s having energy $E(k')$. The elastically backscattered electrons obey the following conservation laws for energy and momentum:

$$E(k') = E(k) \tag{3.5}$$

$$k'_{||} = k_{||} - g \tag{3.6}$$

where $k_{||}$ and $k'_{||}$ are the components of the electron momentum parallel to the surface before and after scattering, respectively, and g is a vector of the reciprocal lattice of the surface (cf. chapter 9). The solution leads to the so-called *intensity pattern*, i.e. the angular distribution of the backscattered intensities at fixed primary energies, and the *intensity profile*, i.e. the variation of I with the primary energy for individual diffraction spots.

In the case of ILEED the conservation laws become:

$$E(k') = E(k) - E_L \tag{3.7}$$

$$k'_{||} = k_{||} - p_{||} - g \tag{3.8}$$

where E_L is the energy loss, which is transferred by the electron in the scattering process, and $p_{||}$ is the corresponding momentum transfer associated with the inelastic scattering process.

For comparison with experiment the solution of the problem has to be formulated in the following terms:

a) the energy profile, for which the directions of the primary and the scattered beams and the energy loss E_L are held fixed and the variation of the intensity I is considered as a function of the primary energy E_p (fig. 3.6a).

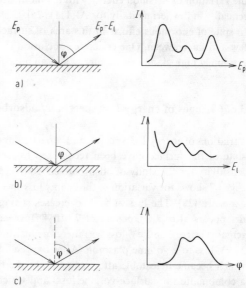

Fig. 3.6. The types of information from ILEED experiments.
a) Energy profile: $I(E_p)$ at constant E_L and φ.
b) Energy loss profile: $I(E_L)$ at constant E_p and φ.
c) Angular profile: $I(\varphi)$ at constant E_p and E_L.

b) the loss profile, where the directions are fixed together with the primary energy E_p, and the variation of the intensity with the energy loss E_L is recorded (fig. 3.6b).

c) the angular profile, for which the intensity is treated as a function of the direction of the scattered beam at fixed direction of the primary beam and constant values for the primary energy E_p and the energy loss E_L.

Duke has shown that, as a good approximation, the problem may be tackled using a two-step model consisting of an elastic diffraction and an energy-loss process. The elastic diffraction may occur either before or after the energy loss (DL- or LD-process). It is clear that in the case of a DL-process a maximum can be expected in the intensity-profile corresponding to a Bragg reflection for the primary energy E_p, whereas a maximum in the same position caused by an LD process requires a primary energy $E_p + E_L$. Effects of this type were observed experimentally more than three decades ago [24].

It appears therefore from the foregoing that something like "inelastic diffraction patterns" should occur, and furthermore that the previously measured energy loss spectra (energy distribution of backscattered electrons at fixed primary energy and integrated over a larger angle) are very sensitive to variations of the primary energy as is observed experimentally.

The question arises whether or not ILEED is of any significance for surface research, since even the theory of elastic LEED intensities is so complicated that for practical purposes no satisfying solution has yet been found. ILEED could probably become a useful technique for determining the electronic structure of valence electrons at the surface, or more precisely, to probe the electronic surface excitation spectra of solids. Duke and coworkers [37] developed a methodology which is mainly based on correlating prominent features in ELEED spectra with those of the ILEED data. If only the loss-profiles are analyzed then the contribution from elastic diffraction may be separated and probably be treated kinematically. Otherwise a complete dynamical theory will be necessary. So far only the surface-plasmon dispersion relation (i. e. the variation of plasmon energy with the momentum transfer $p_{||}$) has been tackled both theoretically and experimentally for Al(111) [25].

In spite of enormous efforts in this area of surface research results have hitherto been rather few and far between. One concrete conclusion is that besides complex theoretical treatments extremely high standards of energy and angular resolution are required of the spectrometer.

3.6. Changes of energy-loss spectra by adsorbed layers

Variations of energy-loss spectra, and in particular of the plasma-loss features, with the chemical state of the surface have been recognized for several years, but only recently has the high sensitivity for the study of adsorbate-covered surfaces been realized.

Fig. 3.7 shows the variation of the energy loss spectrum of titanium as a function of the oxygen exposure [28]. The loss at 5 eV decreases strongly with oxygen adsorption and completely disappears after an exposure of $7 \cdot 10^{-6}$ Torr $\cdot$ s, whereas the intensity of the 12 eV – loss remains large. The 5 eV-loss was therefore attributed to a surface plasmon excitation, and the 12 eV-loss to a volume plasmon. After oxygen adsorption a new loss peak at 24 eV appears, which became detectable after reduction of the background-intensity near 20 eV (caused by a combination of surface-volume loss) and which was suggested to be the second harmonic of the bulk loss.

The growth of a thin oxide layer on Al was studied by Murata and Ohtani [32] by observing the energy shifts of the surface plasmon loss. According to the early theoretical treatment of Stern and Ferrell [14] the dispersion relation of the surface plasmon frequency ω_s in the case of a semi-infinite metal coated with a film with thickness d should be given by

$$\omega_s = \omega_p \left[(\varepsilon + \tanh k d)/(2\varepsilon + [1 + \varepsilon^2] \tanh k d) \right]^{1/2} \tag{3.9}$$

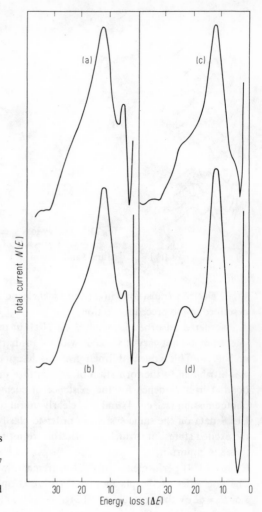

Fig. 3.7. Energy-loss spectra for Ti at various oxygen exposures.
a) clean surface, b) $3 \cdot 10^{-7}$ Torr · s, c) $9 \cdot 10^{-7}$ Torr · s, d) $7 \cdot 10^{-6}$ Torr · s.
Primary energy $E_p = 100$ eV. After Simmons and Scheibner [28].

where ε is the dielectric constant of the oxide film and k is the surface wave vector excited by the incident electron.

Using a primary energy of 350 eV with the beam striking the surface at an angle 45°, and the analyzer collecting electrons scattered at the same angle from the surface, variations of the loss spectrum under continuous gas exposure were obtained as a function of time as shown in fig. 3.8. The energy shift of the surface plasmon loss from 10 eV to 7 eV was used to determine the thickness of the layer in the range up to 30 Å. Although this interpretation is by no means unambiguous it demonstrates potential applications of EELS.

Besides the variation of the intensities and energies of the plasmon losses caused by surface overlayers new peaks in the loss spectrum may also appear after chemisorption. This effect was observed for the first time by Steinrisser and Sickafus [33] and ascribed to two-step

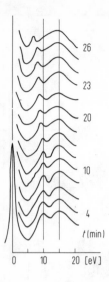

Fig. 3.8. Energy-loss spectra of Al after different durations of oxidation at an oxygen pressure of 10^{-8} Torr. Primary energy $E_p = 350$ eV. After Murata and Ohtani[32].

single electron transitions involving an inelastic loss and elastic backscattering of the type described in the preceding section.

A disordered adsorbed layer of N on Ni(110) revealed an energy distribution whose peak intensities did not vary as strongly with the primary energy as those for an ordered S overlayer on Ni(110). This is due to diffraction effects described by ILEED theory. The origin of the peaks was attributed to the formation of new surface molecular orbitals due to the chemisorption bond. Direct evidence for the existence of such levels has been obtained in particular by photoemission (see ch. 4) and it is clearly useful to compare photoelectron spectroscopy and EELS data on the same system in order to clarify the mechanism. In particular location of the excited states into which the electron from the chemisorption molecular orbital is transferred is important.

Küppers [34] performed similar measurements with CO, H_2 and O_2 adsorbed on Ni(110). The intensities of the plasmon peaks were strongly dependent on the primary energy E_p. For $E_p = 62$ eV the structure of $N(E)$ near the elastic peak is relatively smooth. Fig. 3.9 demonstrates variations of the energy distribution resulting from exposure to CO. These changes were virtually complete after a total exposure of 1.5 L (1 L = 10^{-6} torr · s). The final spectrum shows two chemisorption induced peaks at 5.5 and 13.5 eV below the elastic peak position. At low coverages the higher loss peak occurs at 15 eV. A peak at 15 eV is also seen if the CO-covered sample is heated in UHV to 475 K (curve f). This peak indicates the existence of a second adsorbed species different from molecular CO, which could be either oxygen or carbon since the impinging electron beam can cause dissociation of CO chemisorbed on Ni [35]. Indeed, loss spectra recorded following exposure of the clean surface to oxygen showed a peak at 15 eV. Heating the sample (after spectrum f) in UHV to 775 K resulted in a clean surface as is seen in spectrum g). There is still some uncertainty about the correct interpretation of the observed peaks, which could for example be explained by excitation of electrons from filled to unfilled chemisorption states. In any case this experiment shows the sensitivity of EELS to the state of the surface, providing an electronic "fingerprint" of the adsorbed particles.

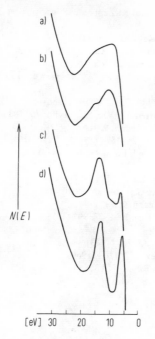
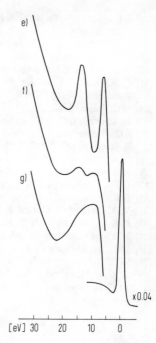

Fig. 3.9. Energy-loss spectra from a Ni(110) surface interacting with CO.
a) clean surface, b) after a CO exposure of $3 \cdot 10^{-7}$ Torr $\cdot$ s, c) $7 \cdot 10^{-7}$ Torr $\cdot$ s, d) $1.5 \cdot 10^{-6}$ Torr $\cdot$ s,
e) $5 \cdot 10^{-6}$ Torr $\cdot$ s, f) after subsequent heating to 200 °C and g) to 500 °C.
Primary energy $E_p = 62$ eV. After [34].

3.7. References

[1] R.L. Gerlach, J.E. Houston and R.L. Park, Appl. Phys. Lett. *16*, 179 (1970).
[2] J.O. Porteus, in: The structure and chemistry of solid surfaces. (G.A. Somorjai, ed.), Wiley, New York 1969, p. 12-1.
[3] C.J. Davisson and L.H. Germer, Phys. Rev. *30*, 705 (1927).
[4] C.C. Chang, Surface Sci. *23*, 283 (1970).
[5] R.L. Robins and J.B. Swan, Proc. Phys. Soc. (London) *76*, 857 (1960).
[6] H.E. Bishop and J.C. Rivière, Appl. Phys. Lett. *16*, 21 (1970).
[7] R.L. Gerlach, in: Electron Spectroscopy (D.A. Shirley, ed.) North Holland, Amsterdam 1972, p. 885.
[8] R.L. Gerlach, J. Vac. Sci. Techn. *8*, 599 (1971).
[9] R.L. Gerlach and A.R. Du Charme, Surface Sci. *29*, 317 (1972).
[10] D. Bohm and D. Pines, Phys. Rev. *85*, 338 (1952); *92*, 609 (1953).
[11] H. Raether, in: Springer Tracts in Modern Physics. Vol. *38* (1965), p. 84, Springer-Verlag Berlin.
[12] D. Pines: Elementary Excitations in Solids. W.A. Benjamin, New York 1963.
[13] R.H. Ritchie, Phys. Rev. *106*, 874 (1957).
[14] E.A. Stern and R.A. Ferrell, Phys. Rev. *120*, 130 (1960).
[15] S. Nozières and D. Pines, Phys. Rev. *113*, 1254 (1959).
[16] E. Rudberg, Phys. Rev. *50*, 138 (1936).
[17] R.A. Ferrell, Phys. Rev. *111*, 1214 (1958).
[18] D. Pines, in: Solid State Physics. (F. Seitz and D. Turnbull, eds.), Vol. 1 (1955), p. 873, Academic Press, New York.

[19] J. Daniels, C. V. Festenberg, H. Raether and K. Zeppenfeld, in: Springer Tracts in Modern Physics. *54*, 77 (1970).
[20] C. B. Duke and G. E. Laramore, Phys. Rev. *B3*, 3183 (1971).
[21] P. I. Feibelman, C. B. Duke and A. Bagchi, Phys. Rev. *B5*, 2436 (1972).
[22] C. B. Duke and U. Landman, Phys. Rev. *B6*, 2956, 2968 (1972).
[23] C. B. Duke, in: Electron Emission Spectroscopy (W. Dekeyser et al., eds.). Reidel Publ. Comp., Dordrecht 1973, p. 1.
[24] J. C. Turnbull and H. E. Farnsworth, Phys. Rev. *54*, 507 (1938).
[25] U. Landman, C. B. Duke and J. O. Porteus, J. Vac. Sci. Techn. *10*, 183 (1973).
[26] J. O. Porteus and W. N. Faith, J. Vac. Sci. Techn. *9*, 1062 (1972).
[27] B. Heimann and J. Hölzl, Phys. Rev. Lett. *26*, 1573 (1971); Z. Naturf. *27a*, 408 (1972).
[28] G. W. Simmons and E. J. Scheibner, J. Appl. Phys. *43*, 693 (1972).
[29] H. Raether, Surface Sci. *8*, 233 (1967).
[30] E. J. Scheibner and L. N. Tharp, Surface Sci. *8*, 247 (1967).
[31] L. K. Jordan and E. J. Scheibner, Surface Sci. *10*, 373 (1968).
[32] Y. Murata and S. Ohtani, J. Vac. Sci. Techn. *9*, 789 (1972).
[33] F. Steinrisser and E. N. Sickafus, Phys. Rev. Lett. *27*, 992 (1971).
[34] J. Küppers, Surface Sci. *36*, 53 (1973).
[35] H. H. Madden and G. Ertl, Surface Sci. *35*, 211 (1973).
[36] W. H. Weber and M. B. Webb, Phys. Rev. *177*, 1103 (1969).
[37] a) C. B. Duke and U. Landman, Phys. Rev. *B 7*, 1368 (1973); *B 8*, 505 (1973).
     b) C. B. Duke and A. Bagchi, J. Vac. Sci. Techn. *9*, 738 (1972).

4. Photoelectron Spectroscopy

4.1. Introduction

Some twenty years ago K. Siegbahn in Uppsala began intensive efforts to improve the energy resolution and sensitivity of β-spectrometers to allow measurements of the electronic binding energies in atoms. The principle of the experiment is simple: electrons are liberated from their bound states with energies E_b by interaction with electromagnetic radiation, and the kinetic energy of these electrons E_{kin} which is then measured is given by

$$E_{kin} = h v - E_b \tag{4.1}$$

where v is the known frequency of the exciting radiation. In Siegbahn's work soft x-ray sources with energies in the range of about 1 keV were used. Siegbahn developed his spectrometer to the extent that even "chemical" shifts in the electron binding energies of the order of 1 eV and less became detectable, and he named the technique "electron spectroscopy for chemical analysis" or ESCA accordingly. The work of this group has been reviewed in two monographs [1, 2].

A somewhat different method of determining ionization potentials was used by Turner [3], who performed the excitation by means of He resonance radiation (21.2 eV) and selected the emitted photoelectrons with a retarding field analyzer. This technique is restricted mainly to electrons in the outer shells of atoms and molecules and has become an invaluable tool in molecular spectroscopy [4, 5]. The solid state physicist first became interested in the energy distribution of photoelectrons from semiconductors to obtain information about electronic band structure [6]. No detailed data on the escape depths of slow electrons from solids existed at this time and the results were considered as reflecting bulk properties, which was essentially true, since the energy of the primary radiation was below 10 eV (compare fig. 1.2). In 1964 Berglund and Spicer [7] reported photoemission data for Cu which were found to be in agreement with the known electronic properties (i. e. density of states) of this material. This initiated a series of investigations in particular with transition metals [8]. In all experiments the energy of the exciting electromagnetic radiation was below 11.6 eV, which is the cut-off energy of LiF-windows.

The sensitivity of this "ultraviolet photoelectron spectroscopy" (UPS) to surface properties was at first clearly demonstrated by Eastman [9] for oxygen adsorbed on nickel. Eastman and his coworkers [10] further showed that the primary energy of radiation must be at least 20 eV in order to obtain "true" information about the electronic energy distribution in the solid and at its surface. With lower energies complications arise from the final densities of states, apart from the fact that the method becomes less surface – sensitive for electron energies < 20 eV due to their large escape depth. In 1971 Eastman and Cashion [11] were able to correlate observed structure in the energy distribution curve of CO covered Ni with the electron energies in free CO and with the formation of chemisorption orbitals.

"X-ray photoelectron spectroscopy" (XPS or ESCA) differs from UPS only in the energy of the primary radiation (~ 1 keV compared with 20–40 eV in UPS), the instrumentation being essentially the same. Currently attempts are being made to bridge the gap between XPS and UPS by using x-ray sources with lower characteristic radiation energies and by increasing the UV energy, e. g. by the use of synchrotron radiation. Detailed information about the various fields of application of photoelectron spectroscopy as well as of the present state of the art can be found in the proceedings of a recent conference [13]. Photoelectron spectroscopy is one of the exciting and rapidly growing branches in chemistry and solid state physics. As an example of its application to free molecules fig. 4.1 shows the UPS spectrum of CO.

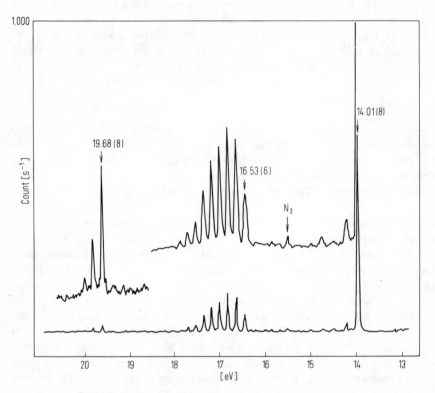

Fig. 4.1. UPS spectrum of free CO molecules (Turner et al. [5]).

Photoelectron spectroscopy is also useful in surface research, although this has been recognized only recently. According to fig. 1.2 the depth of information is about 6–8 Å with UPS and about 20–30 Å with XPS. In principle the following applications are possible:

a) Chemical analysis of the surface from the derivation of the binding energies of core electrons.
b) Information about the atomic charge from "chemical shifts" of the binding energies of core electrons.
c) Determination of the density of states of the valence electrons.

Method a) applies only to XPS, and although no systematic investigations have been made, the available data [62] indicate clearly that this method is able to detect surface concentrations in the sub-monolayer region, but is much less sensitive than electron stimulated Auger electron spectroscopy. XPS requires more experimental refinement than AES so that no widespread application just for the chemical analysis of surfaces is to be expected. (The situation may well be somewhat different if one is interested in the composition of a region near the surface with about 20 Å thickness). The primary x-rays also excite Auger electrons which are recorded in the XPS spectra, but again the use of normal Auger spectroscopy (i.e. using an electron beam) appears to be more favourable for surface work.

"Chemical shift" denotes the effect, in which the binding energy of a core electron is influenced by electrostatic interaction with the valence electrons. The shifts may be as large as a few eV and give information about the atomic charge or the degree of ionicity (see for example [13]). These phenomena are of particular interest with ions of transition metals which have several oxidation states [63]. These effects can also be studied by AES, but there the interpretation is more complex since it deals with a two-electron process in contrast to XPS where only a single electron is involved. The results for atoms in the surface monolayer are so far rather scant [65]. The most exciting aspect for surface research is the possibility of studying spectra from valence electrons by means of UPS, a technique which is more surface sensitive than XPS, and the following discussion concentrates mainly on this topic.

4.2. Experimental

The experimental arrangement for photoelectron spectroscopy at solid surfaces is shown schematically in fig. 4.2.

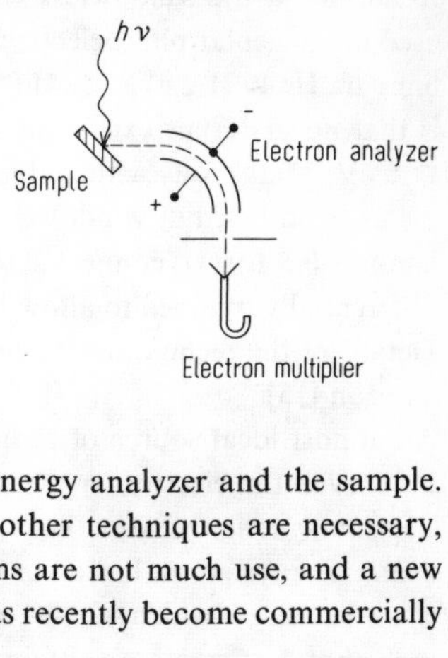

Fig. 4.2. Schematic diagram of the experimental arrangement for photoelectron spectroscopy of solids.

The essential components are the light source, the electron energy analyzer and the sample. For surface studies UHV conditions and combination with other techniques are necessary, so that earlier ESCA spectrometers with poor vacuum systems are not much use, and a new generation of photoelectron spectrometers for surface work has recently become commercially available.

4.2.1. Light sources

X-ray photoelectron spectra are obtained using x-ray tubes with different anode materials providing characteristic radiation on a bremsstrahlung background. Hitherto the Al-K_α (1486.6 eV) and Mg-K_α (1253.6 eV) radiation have proved most popular. Both sources may be combined as a switchable twin-anode which enables the separation of photoemitted electrons (photolines) whose kinetic energy depends on the photon energy, and of Auger electrons, whose energy is fixed. An Yttrium source [12] which provides characteristic radiation at 132 eV is an attractive proposition for surface studies, but needs some further development.

The main disadvantage of these x-ray sources is the inherent energy width of the lines, which is about 1 eV in the case of Al and Mg. The energy resolution is thus limited at least to this value although the energy analyzers have far better properties. Some improvement is offered by x-ray monochromators which operate by diffraction giving a resolution of 0.5 eV which will probably be reduced to 0.2 eV in the future. The natural width of the core levels of light atoms, such as C 1s, in free molecules are about 0.1 eV and upon which chemical shifts up to 10 eV have been observed. However x-ray monochromators are expensive and have only a relatively low transparency, so that the already low photon yields from the sample are reduced further. The use of K_α radiation from a Na-source (energy 1041 eV) [64] seems quite promising since this line has an energy width of only 0.6 eV, so that for most purposes a monochromator will probably not be necessary.

In order to prevent gas from streaming from the x-ray tube into the analyzer chamber it is frequently isolated from the main system by a thin window (C, Be) and pumped separately. Gas discharge lamps are used in the ultraviolet region. They consist of a tube filled to a few hundred microns pressure of the working gas and containing two metal electrodes between which a voltage of a few kilovolts is applied. Depending on the particular experimental conditions either discrete line spectra or continuous spectra are produced. In the continuum mode a UV monochromator is used to select the desired wavelengths. In earlier studies [8] energies below 11.6 eV (the cutoff energy of LiF windows) were used produced by a H_2 gas discharge. In surface work windowless spectrometers [14] using a shorter wavelength source such as used in molecular photoelectron spectroscopy [5] are preferable. The most suitable resonance lines are He I (21.2 eV), He II (40.8 eV), Ne I (16.7 eV), and Ne II (26.9 eV). A major problem is that no materials exist which are transparent to UV radiation with energies greater than 11.6 eV. Metal foils with a thickness of only a few hundred Ångstroms are of some limited use as windows, but windowless spectrometers are normally used. In this case the discharge lamp (~ 0.5 Torr) is connected with the analyzer chamber by a capillary or capillaries which are differentially pumped to allow UHV conditions to be maintained in the sample chamber [14]. Details of the techniques for producing and operating UV radiation have been described by Samson [15].

An almost ideal source of radiation covering the whole energy range between 10 and a few 1000 eV is the electron synchrotron, in which relativistic electrons in a ring accelerator emit electromagnetic radiation into a very narrow solid angle tangential to the circulating electron beam. The radiation has a continuous spectral distribution whose maximum depends on the electron energy (which is in the range of about 1 GeV). Since this collimated beam is very intense high photon currents are obtained even at the exit slit of an UV-monochromator. An additional property of the synchrotron radiation is that the emitted photons are linearly polarized. Unfortunately this technique is limited to relatively few research groups for obvious reasons.

4.2.2. Analyzer and detector

The spectrometers used by Siegbahn in his earlier work used magnetic fields to disperse the electrons for energy analysis [1]. These instruments have excellent focusing properties but a major disadvantage is the non-linear relation between the magnetic field strength and the selected electron energy, and nowadays electrostatic energy analyzers are normally used. XPS instruments of the present generation are most often based on a concentric hemisphere analyzer together with a retarding lens system which decelerates the incoming electrons in order to improve the resolution, and the same type of analyzer may also be used in UPS work. Several other types are also suitable for photoelectron spectroscopy, and cylindrical electrostatic analyzers [14, 18], retarding field analyzers [19] (which can be improved by means of a 'post-monochromator' collector [16]), and cylindrical mirror analyzers [17, 20] have been used successfully.

An electron multiplier such as a channeltron or a channelplate [12] is useful to detect the energy selected electrons. The relatively small intensities produced in photoelectron spectroscopy frequently require refined handling of the data. Commercially available spectrometers are therefore equipped with a signal averager or are compatible with such instruments. Phase sensitive detection of the signal is unsuitable for use with a channeltron or a channelplate.

The spectrometer is scanned by varying the retarding voltage between the sample and the entrance slit (analyzer with pre-retardation), the electrode potentials of the lens (analyzer with retarding lens) or the potentials applied to the analyzer electrodes themselves. It is therefore relatively simple to average the data digitally over many consecutive spectra by controlling the scan with the sweep generator of a multichannel analyzer.

If a channelplate multiplier is used as detector an interesting possibility is offered for extending the "point by point" recording procedure. The channelplate is placed behind the exit slit so that the individual channels subdivide the energy range passing through the slit even further, giving the effect of making many individual high resolution measurements at the same time. Depending on the total number of amplifying channels this technique yields a considerable improvement over conventional analysis.

Semi-automatic operation of some commercial spectrometers is possible where the sampling and recording of the spectra and the changing of samples is controlled by a small computer. As already outlined, in XPS work the energy resolution is always limited by the line width of the source. With UPS much better resolution is obtainable down to $10-50$ meV, which allows individual vibrational states in free molecules to be detected. The practical limitations result from minor potential variations within the spectrometer at the slit and electrodes. In surface work resolution of 0.1 eV has proved quite adequate so far, since features in the energy distribution curves (EDC) from solid surfaces requiring higher resolution have not been observed. One further important fact concerns the origin of the energy scale of the measured energy distribution. Equ. (4.1) is incomplete in that the work function of the analyzer electrodes ϕ_{sp} (but not that of the sample!) is not included. This is necessary because the analyzer selects the electrons with respect to the vacuum level, as illustrated by fig. 4.3, so that equ. (4.1) must be replaced by

$$E_{kin} = h\nu - E_b - e\phi_{sp} \qquad (4.2)$$

Thus to determine electron binding energies with respect to the vacuum level the work function ϕ_{sp} must be known which is usually not the case, so that in studies with free molecules a reference is needed.

In work on metal surfaces (which is the main concern here) the origin of the energy scale is taken as the Fermi level E_F which can be located easily in the EDC since electron emission sets in at this energy. If this energy scale is to be referred to the vacuum level (e. g. when chemi-

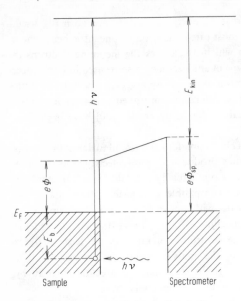

Fig. 4.3. Energy level diagram for photoelectron spectroscopy. A photon of energy $h\nu$ is absorbed by an electron at a level E_b below the Fermi level E_F. The measured kinetic energy E_{kin} of the ejected electron is given by eq. (4.2).

sorption states are compared with the levels of the corresponding free molecules) then of course the work function of the sample $e\phi = E_{vac} - E_F$ is needed.

With semiconductors the onset of the emission does not give the position of the Fermi level but that of the valence band edge. The position of the Fermi level may be located by studying the photoelectron emission from a thin metallic layer evaporated onto the semiconductors [20]. Additional effects due to band-bending in the semiconductors surface layer must also be considered.

Location of the Fermi level of oxides and insulators [22] is more complicated still, since charging effects must also be taken into account.

4.3. Photoelectron emission from solids

The kinetic energies of photoemitted electrons are related to the binding energies of electrons in molecules or solids according to equ. (4.2). A major complication in the study of solids is that the excited electrons may suffer inelastic effects in the bulk and can give rise to the emission of secondary electrons, plasmon excitations etc. This effect can best be estimated by looking at a core level near the valence band. A typical result of these effects is to produce a flat background, which is also present in the energy distribution curves from bands and can hence be subtracted.

There are no further special problems inherent to the solid state if the core states of the atoms are measured. As an example fig. 4.4 shows a spectrum from stainless steel taken with XPS. Since even electrons with energies of 1 000 eV originate from depths of less than 20 Å a considerable part of the detected photoelectrons is due to excitations of surface atoms, so that this technique can be readily used to characterize the chemical composition of the surface by identifying the core level energies. In addition XPS spectra contain Auger electrons which can also be used for chemical analysis. However common Auger electron spectroscopy

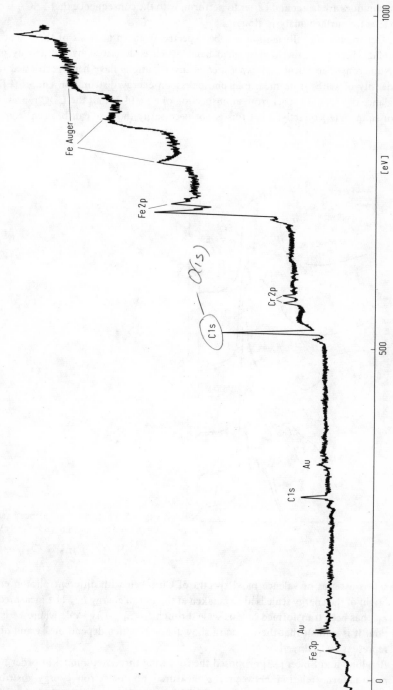

Fig. 4.4. XPS spectrum of stainless steel after heating to 450 °C under UHV conditions. Light source: Al K_α, non monochromatized (Courtesy of Vacuum Generators. Ltd.).

techniques are faster and easier to perform, with the consequence that ESCA is only of limited value for surface analysis alone.

Most interesting information can be expected from an analysis of the valence electrons. In solids these form the highest filled band which is described by the density of states $N(E)$. Numerous theoretical and experimental investigations have been performed to evaluate the density of states from measured photoelectron energy distribution curves (EDC). At a first glance one would expect from an inspection of equ. (4.2) that the EDC represents the density of states $N(E)$ directly. That this is not necessarily the case can be seen from fig. 4.5 which

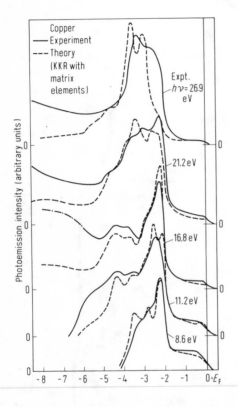

Fig. 4.5. Photoelectron valence band spectra for Cu for photon energies $8.6 \leq h\nu \leq 26.9$ eV (Eastman[25]).

shows a series of valence band spectra of Cu taken with different photon energies $h\nu$. The origin of the energy scale is always taken at the Fermi energy E_F. The measured kinetic energy E_{kin} has been transformed on this scale through $E = E_{kin} - h\nu + e\phi$, where $e\phi$ is the work function. It is evident that these spectra show features which depend on $h\nu$ and obviously do not represent $N(E)$ directly.

Berglund and Spicer [23] proposed the following three-step model in order to obtain an approximate correlation between the measured photoelectron energy distribution and the density of electronic states in the solid. This model is by no means able to yield a completely correct description, for which more complicated theoretical approaches are necessary [40].

(a) Excitation of an electron in the solid from the level E_i to the level $E_j = E_i + h\nu$ by means of absorption of a photon with energy $h\nu$.

(b) Propagation of the excited electron to the surface of the solid.

(c) Ejection from the solid into the vacuum.

If for the moment one assumes that the internally produced density of electrons is not modified by steps (2) and (3), then those electrons whose energies exceed the work function will be able to leave the solid. The EDC then depends on the densities of states $N(E_i)$ and $N(E_j)$ (the states E_j being empty) and on the probabilities for the transitions $E_i \rightarrow E_j$ at a given energy $h\nu$ of the radiation. The first task therefore is to evaluate this "internal" distribution of excited electrons $N_{int}(E)$.

The state of an electron within an energy band of a solid is characterized by its energy E_i and its momentum k_i. If the electron is excited to an energy E_j, then in principle the momentum may be conserved, i.e. $k_j = k_i$ (direct transitions) or it may change, $k_j \neq k_i$ (indirect transitions). One important conclusion from both experimental and theoretical work is that k-conservation normally holds, which means that in the main only direct transitions occur in photoelectron emission [24]. Complete information about the electronic states in a solid is contained in the $E(k)$ relation (which can be determined experimentally in the region near the Fermi level by de Haas – van Alphen experiments). The density of states $N(E)$ is an integration in momentum space over all states with equal energy but different k – values within a certain band, and summed over all bands n [25]:

$$N(E) = \frac{1}{(2\pi)^3} \sum_n \int d^3k \, \delta\left[E - E_n(k)\right] \tag{4.3}$$

It can be assumed that the number of electrons excited from E_i to E_j is proportional to the density of filled states at E_i, $N(E_i)$, (where $E_i < E_F$) and proportional to the density of unoccupied states at E_j, $N(E_j)$, (where $E_j = E_i + h\nu > E_F$), multiplied by the transition probabilities (or interband momentum matrix elements) $p_{i,j}(k)$.

In terms of equ. (4.3) this leads to the relation

$$N_{int}(E, h\nu) = C \cdot \sum_{i,j} \int d^3k \, |p_{i,j}|^2 \, \delta(E_j - E_i - h\nu) \cdot \delta(E - E_i) \tag{4.4}$$

where C is a normalization factor.

All possibilities for an electron to be excited from band $E_i(k)$ to band $E_j(k)$ (with conservation of the momentum) are summed, but only those final states E_j which are at energy E are selected. The total ensemble of transitions between E_i and E_j (whereby $E_j = E_i + h\nu$) is called the "joint optical density of states" (JODS) $D(h\nu)$:

$$D(h\nu) = \frac{1}{(2\pi)^3} \cdot \sum_{i,j} \int d^3k \, \delta[E_j(k) - E_i(k) - h\nu] \tag{4.5}$$

where

$$E_i < E_F < E_j$$

Obviously the internal distribution $N_{int}(E, h\nu)$ is closely related to the JODS $D(h\nu)$. $N_{int}(E, h\nu)$ contains just those joint densities which lead to the energy E (weighted by the matrix element p_{ij}).

In order to correlate the internal distribution $N_{\text{int}}(E, h\nu)$ with the measurable external energy distribution $N_{\text{ext}}(E, h\nu)$ the next step is to consider the propagation of the excited electrons to the surface. Excitation of electrons takes place at a maximum depth κ, which represents the extent to which the primary radiation penetrates the solid. κ depends on the wavelength and on the optical constants of the material and is of the order of 100 Å for UV radiation and even more for x-radiation. If the energy of the excited electrons exceeds about 8–10 eV their mean free path in the solid is therefore much smaller than the optical path κ. On their way to the surface the photoelectrons can produce cascades of secondary electrons whose concentration increases towards lower kinetic energies. This effect causes an apparent increase in the EDC with increasing binding energy.

Before escaping into vacuum an electron has to surmount the surface barrier presented by the work function $e\phi$ which in the free electron approximation is equivalent to the condition that the component k_z of the momentum normal to the surface must exceed a critical value [23]. More generally an escape function $T(E_j, \mathbf{k})$ is introduced which describes the probability for an excited electron in state $E_j(\mathbf{k})$ to reach the surface and leave the solid [24], thus

$$N_{\text{ext}}(E, h\nu) = N_{\text{int}}(E, h\nu) \cdot T(E_j, \mathbf{k}) \tag{4.6}$$

Equ. (4.4) then becomes

$$N_{\text{ext}}(E, h\nu) = C \sum_{j,j} \int d^3 \mathbf{k} |p_{i,j}|^2 \, \delta \, [E_j(\mathbf{k}) - E_i(\mathbf{k}) - h\nu] \cdot \delta \, [E - E_i(\mathbf{k})] \cdot T(E_j, \mathbf{k}) \tag{4.7}$$

Using this formula N_{ext} can be calculated (at least in principle) and compared with experimental results which has been done in some cases.

To a first approximation the transmission function $T(E_j, \mathbf{k})$ is only important near the threshold and with lower photon energies. If the unknown matrix elements are neglected, the measured EDC $N_{\text{ext}}(E, h\nu)$ represents more the joint density of states (i. e. a combination of occupied and unoccupied states) than the occupied band density of states which is of primary interest. The variation of the structure of the EDC's in fig. 4.5 with changing $h\nu$ is interpreted as variations in the band structure of the non-occupied states. However the EDC's also contain the important features of $N(E)$. A strong increase at about 2 eV below E_F obviously represents the position of the upper limit of the d-band, whose width can be estimated to be about 5.7 eV. With $h\nu > 22$ eV the overall character of the EDC's becomes less dependent on $h\nu$. It can therefore be concluded that at primary photon energies exceeding 20 eV (or still better 40 eV) the measured energy distributions represent the density of occupied states $N(E)$ fairly well. At higher photon energies the final densities of highly excited electrons have negligeable structure, and UPS results become very similar to XPS spectra of the valence bands and furthermore agree with calculated band structures.

However experience shows that a lot of information may also be obtained from UPS spectra using He(I) (=21.2 eV) or lower energy light sources for excitation. Occupied band widths and positions as well as the formation of chemisorption induced states for example may be observed in many cases directly from the spectra. More quantitative conclusions can be arrived at by averaging a set of EDC's recorded at different photon energies, as proposed by Eastman [25].

There remain some open questions in photoelectron spectroscopy, such as the unknown parameters of matrix element modulation, the influence of hole lifetime on the spectrum, and the role of collective effects, for example polarization of the environment of the hole. At lower photon energies, apart from the structure of the unoccupied states the surface transmission coefficient, which is poorly understood and which becomes important near threshold, and the secondary electron emission are disturbing factors.

For surface research photon energies around 40 eV would appear to be most favourable since electrons within this energy range have minimum escape depth and therefore the technique should be very surface sensitive. XPS measurements are also useful, although the escape depths are larger and the energy resolution somewhat poorer.

4.4. Band structure of metals and alloys

The main field of application of photoelectron spectroscopy to solid state problems is the investigation of the electronic structure of metals and alloys. Although these are primarily bulk properties we know that this method probes the surface region. No special effects have been detected so far which have to be ascribed uniquely to electronic states at the (clean) surfaces of metals – with one exception. This leads to the belief that for metals the electronic density of states at the surface is essentially identical to that of the bulk. A knowledge of these properties is of vital importance in attempts to understand the formation of chemical bonds between metal surfaces and adsorbed particles, and a somewhat more detailed discussion of this question is given in the next section.

As an example fig. 4.6 shows the valence band spectrum of Au as obtained by Shirley [27]

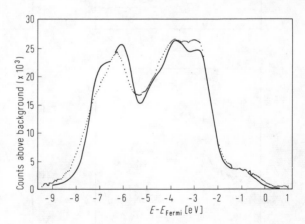

Fig. 4.6. XPS valence band spectrum of Au (points) together with the results of a calculation by Connolly. The spectrum was obtained with monochromatized Al K_α-radiation (Shirley [27]).

by XPS using an x-ray monochromator. The agreement with the theoretical density of states is fairly good, which justifies considering the EDC as being a more or less direct measure of $N(E)$. Numerous other spectra (UPS and XPS) from the valence bands of metals are to be found in the literature [8], [25]–[29].

The results demonstrate that the three step mechanism as outlined in the preceding section yields an essentially adequate description of the process and that the main problems arise from the effect of the matrix elements and of the final densities of unoccupied states. A theoretical evaluation of the EDC's of Cu has been performed by Williams et al. [30]. Their curves exhibit structures varying with hv as found by experiment (fig. 4.5).

Various attempts have been made to measure variations in the electronic structure of Ni with temperature. Whereas Pierce and Spicer [31] reported a negative result, Rowe and Tracy [32] observed some slight anomalous critical behaviour in the d-band peak energy vs. temperature in the vicinity of the Curie temperature.

The investigation of the electronic structure of binary alloys is at present an area of much experimental and theoretical activity. It has been found that for systems like Cu/Ni and Ag/Pd the rigid band model is no adequate description but must be replaced by models like the coherent potential approximation [33] which leaves the positions of the d-band in the alloys roughly at the same position as in the pure metals. XPS investigations supporting this idea have been performed by Hüfner et al. [34], [35]. As an example fig. 4.7 shows some spectra

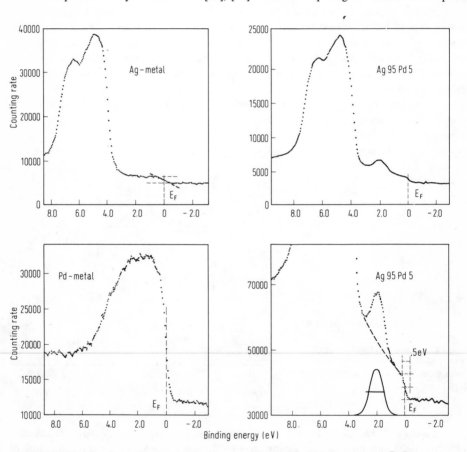

Fig. 4.7. XPS valence band spectra of Ag, Pd, and an Ag/5% Pd alloy (Hüfner et al. [35]).

from Pd, Ag and an alloy containing 5% Pd. In the last the formation of a narrow Pd d-band on top of the Ag sp band can be seen, which is consistent with the formation of "virtual bound states" as formulated by Friedel [36] and Anderson [37] for dilute alloys.

More refined investigations with metals are for example concerned with the angular variation of the photoemission [38] or the emission of spin polarized electrons from ferromagnetic materials [39]. The theory of photoemission must be extended to include these effects. Several attempts to replace the three step mechanism by more refined models have already been made [40].

4.5. Metal surfaces and chemisorption levels

Photoelectron spectra from metals exhibit in general fairly good agreement with the densities of occupied states in the bulk, and one is therefore safe in the assumption that the electronic properties of the surface do not differ significantly from bulk properties. A more detailed inspection however reveals that differences should exist even if only to a minor extent. Haydock et al. [41] concluded from Eastman's UPS data for Cu that the width of the d-band at the surface should be somewhat narrower than in the bulk, as is also implied by data from ion neutralization spectroscopy [42] which is a particularly surface specific technique. The result of theoretical analysis [41, 43] is that the d-band width varies as $\sqrt{z}$, where z is the number of nearest neighbours. In the case of an fcc crystal $z = 7, 8, 9$ for (110), (100) and (111) surfaces, compared with $z = 12$ in the bulk. It remains to be seen whether this phenomenon has a strong influence on the variation with surface orientation of the strength of chemical surface bonds. A second important factor concerns the possible existence of electronic surface states at metals. These are states which are localized at the surface and decay into the bulk and arise from the change of the periodicity and the potential at the surface. These surfaces states have been discussed extensively for semiconductors, but only recently have they been predicted theoretically for metals [44] and used to explain the occurrence of some unusual structure in earlier photoelectron spectra of Cu and Ni. However it was later demonstrated that these peaks were not due to intrinsic surface states but to the presence of surface contaminants [45]. The most important application of photoemission to surfaces lies in the detection of chemisorption levels. These are energy states which are created at the surface by the formation of the chemisorption bond between orbitals of the solid and of the adsorbed particle, and the knowledge of the energies is of vital importance to the understanding of the nature of chemisorption. Some earlier observations on variations of photoemission yields were ascribed to the existence of adsorbed layers [50, 51]. The first conclusive experimental verifications were performed by Eastman and Cashion [11] for CO and oxygen chemisorbed on Ni, using He(I) and He(II) excitation sources. Some of the spectra are shown in fig. 4.8. The extra intensity due to emission from adsorbate states reaches values of about 1/3 of the intensity from the substrate. This value is consistent with the mean free path data for electrons within this energy range which should be about 10 Å, i.e. 1/3 to 1/4 of the total emitted intensity should originate from the first atomic layer. UPS spectra from CO covered Ni (fig. 4.8) indicate the existence of new states from which electron emission is detected at energies of about 7.5 and 10.7 eV below the Fermi level. These features can be correlated with orbitals of the free CO molecule shifted in energy by the formation of the chemisorption bond and broadened because the electrons in these virtual bound states exchange rapidly with valence electrons in the metal. It has to be noted however that the observed positions of the emission peaks do not reflect directly the energies of the chemisorption levels since the former are influenced by work function and relaxation effects to an extent which is still widely unknown.

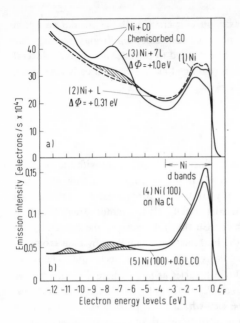

Fig. 4.8. UPS spectra from clean and CO covered Ni surfaces. a) $h\nu = 21.2$ eV; b) $h\nu = 40.8$ eV (Eastman and Cashion [11]).

It is well known that oxygen adsorption on nickel is followed by the formation of oxide with continuing gas exposure. The EDC's in fig. 4.9 [11] illustrate the variation of the electronic

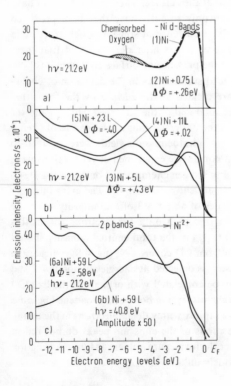

Fig. 4.9. UPS spectra for Ni after interaction with oxygen (Eastman and Cashion [11]).

structure of the solid which is associated with this process. At first a peak at about 6 eV below E_F develops which is accompanied by an increase of the work function and is due to adsorbed oxygen. At increasing oxygen exposure the oxygen-induced levels increase in magnitude and broaden and the work function and Ni d-band emission within 3 eV below E_F decrease. Finally a new peak at 2 eV below E_F appears which is attributed to the d electrons of oxidized nickel and emission in the broad band between 2.5 and 11 eV below, which is ascribed to oxygen 2p-bands.

Similar observations have been made by Helms and Spicer [56] and by Kress and Lapeyre [55] for the interaction of oxygen with Ba and Sr. In contrast to the results with Ni [11] adsorbed oxygen forms rather narrow states in these cases. Again the formation of bulk oxides was associated with characteristic changes of the spectra.

Further results with adsorbate covered surfaces have been obtained by Eastman with H_2, N_2, CO, CO_2 and O_2 adsorbed on Ti[53]. From the spectra it was concluded that CO and CO_2 appear to dissociate into oxygen and carbon. The data obtained with hydrogen indicate the formation of a metallic Ti/H (bulk) compound, where roughly half of the Ti d-electrons are transferred to the hydrogen-originated valence bands.

Similar investigations have been performed with the palladium/hydrogen system [54], where the β-phase PdH exhibits a new band of states centered at 5.4 eV below E_F. These results shed new light on the problem of the electron structure of this system and showed that the widely accepted "proton model" is somewhat misleading. The protons are screened by the electrons in the low-lying Pd-H bonding states. It is concluded that hydrogen is more electronegative than palladium in β-phase PdH. This might also explain the fact that hydrogen atoms adsorbed on Pd and Ni increase the work functions, i.e. become negatively charged.

A detailed explanation of the chemisorption levels discovered is needed. Their common feature is that they are located on the energy scale between 0 and 10 eV below the Fermi level, which indicates that as with all types of chemical bonds the bond formation takes place via interactions between the highest filled and lowest unfilled orbitals of the constituents. Certainly the results obtainable by photoelectron spectroscopy will lend a strong impetus to the theory of chemisorption. A theoretical approach for UPS from adsorbate covered surfaces was developed by Penn [52], which suggests that caution is necessary in interpreting the UPS yields simply as measuring the density of adsorbate states. An antiresonance contribution to the emission current may shift the adsorbate induced peaks to somewhat lower energies and make the line-shape asymmetric. This and in particular the above mentioned relaxation effects must be investigated in the future as a matter of priority. UPS is certainly invaluably useful in identifying the chemical nature of adsorbed particles (e.g. to give an answer whether adsorbed CO dissociates or not). Without any detailed analysis of the data this technique may be used to obtain a "fingerprint" of the chemical (i.e. molecular) state of the adsorbate layer.

The first experimental evidence from photoemission experiments of the existence of "true" surface states (i.e. at clean surfaces) in metals was found by Waclawski and Plummer [19] and by Feuerbacher and Fitton [18] with W(100). Fig. 4.10 shows spectra from clean and CO-covered surfaces. The strong decay of the peak with maximum ~ 0.4 eV below E_F upon gas adsorption is clearly detectable. Simultaneously a new peak at -2.5 eV appears. The existence of these states had been postulated previously by Plummer and Young [48] from their field emission energy distribution curves. These features have only been observed so far with the W(100) surface and not with other surface orientations, and are interpreted as arising from an intrinsic surface state existing in a spin-orbit split gap in the $\langle 100 \rangle$ direction of tungsten [49].

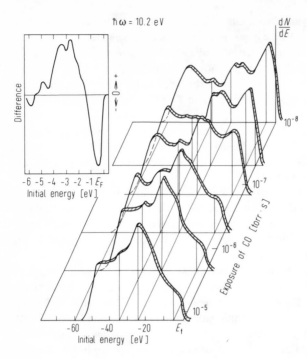

Fig. 4.10. UPS spectra ($h\nu = 10.2$ eV) from a tungsten surface with increasing CO exposure. As the intrinsic surface state at -0.4 eV disappears a new state at -2.5 eV increases in intensity. (Waclawski and Plummer [19]).

In the case of hydrogen adsorption Feuerbacher and Fitton [18] concluded that the surface state is suppressed by particles adsorbed in the β-state. If it is assumed that each adsorbed particle compensates one surface state then a density for the clean surface of about 2.5×10^{14} cm^{-2} (i.e. one surface state per 4 surface atoms) results. It is interesting to note that the W(100) surface is also unique in so far as the work function changes after adsorption of H_2 and N_2 differ considerably from the behaviour of other crystallographic planes.

4.6. Semiconductors

Numerous investigations of the valence bands of semiconductors have been performed in the past by means of UPS at less than 11.6 eV [8a]. A series of recent investigations with higher radiation energy were in very good agreement with the theoretical densities of states. This was so for XPS measurements with Si, Ge, GaAs and ZnSe [46, 47] as well as for UPS with Si, Ge and GaAs using synchrotron radiation [57].

'Intrinsic' surface states at semiconductors are of very great practical importance in semiconductor technology and therefore much experimental effort with indirect methods has already been expended on this field [58]. Direct experimental evidence for the existence of surface states in Si, Ge and GaAs has been found using UPS. On the cleaved (111) face of n-type Si Wagner and Spicer [59] discovered a 1.8 eV wide surface band with a peak at 1.1 eV below E_F containing approximately one electron per surface atom. Essentially similar results were found by Eastman and Grobman [60] for Si, Ge and GaAs. The surface nature of these emission maxima was confirmed by their disappearance after gas exposure.

Certainly such measurements should be combined with other techniques, since each is sensitive to different aspects of the surface states.

Smith and Huchital [61] were the first to use XPS as a monitor for adsorption kinetics by studying the system Cs/GaAs. They assumed that the height of the Cs photoelectron peak is proportional to the surface concentration (below monolayer levels) and concluded from their results that the adsorption kinetics obey the Langmuir model, which postulates that impinging atoms adsorb only on bare surface sites.

4.7. References

[1] K. Siegbahn, C. Nordling, A. Fahlmann, R. Nordberg, K. Hamrin, J. Hedman, G. Johansson, T. Bergmark, S. Karlsson, I. Lindgren and B. Lindberg: ESCA. Atomic, Molecular and Solid State Structure Studied by means of Electron Spectroscopy. Almquist & Wiksells, Uppsala 1967.

[2] K. Siegbahn, C. Nordling, G. Johansson, J. Hedman, P.F. Heden, K. Hamrin, U. Gelius, T. Bergmark, L.O. Werme, R. Manne and Y. Baer: ESCA Applied to Free Molecules. North Holland, Amsterdam 1969.

[3] D.W. Turner and M.I. AlJoboury, J. Chem. Phys. *37*, 3007 (1962).

[4] D.W. Turner: Molecular Photoelectron Spectroscopy. In: Physical Methods in Advanced Inorganic Chemistry. (Eds. H.A.O. Hill and P. Day), Wiley, New York 1968.

[5] D.W. Turner, C. Baker, A.D. Baker and C.R. Brundle: Molecular Photoelectron Spectroscopy. Wiley, New York 1970.

[6] a) G.W. Gobeli and F.G. Allen, Phys. Rev. *127*, 141 (1962).
    b) W.E. Spicer and R.E. Simon, Phys. Rev. Lett. *9*, 385 (1962).
    c) E.O. Kane, Phys. Rev. *127*, 131 (1962).

[7] W.E. Spicer and C.N. Berglund, Phys. Rev. Lett. *12*, 9 (1964).

[8] Reviews:
    a) W.E. Spicer: Photoelectric Emission; in: Optical Properties of Solids. (Ed. F. Abelès) North Holland, Amsterdam 1972.
    b) N.V. Smith: Photoemission Properties of Metals. CRC Critical Reviews in Solid State Sciences, 1971.
    c) D.E. Eastman: Photoemission Spectroscopy of Metals. In: Techniques of Metals Research. (Ed. E. Passaglia.) Vol. *6*, Wiley, New York 1971.

[9] D.E. Eastman, Phys. Rev. *B3*, 1796 (1971).

[10] a) D.E. Eastman and J.K. Cashion, Phys. Rev. Lett. *24*, 310 (1970).
     b) D.E. Eastman and W.D. Grobman, Phys. Rev. Lett. *28*, 1327 (1972).

[11] D.E. Eastman and J.K. Cashion, Phys. Rev. Lett. *27*, 1520 (1971).

[12] M.O. Krause, Chem. Phys. Lett. *10*, 65 (1971).

[13] D.A. Shirley, ed.: Electron spectroscopy. Proc. of the Int. Conf., Asilomar (Calif.) Sept. 1971, North Holland, Amsterdam 1972.

[14] J.K. Cashion, J.L. Mees, D.E. Eastman, J.A. Simpson and E.C. Kuyatt, Rev. Sci. Instr. *42*, 1670 (1971).

[15] I.A.R. Samson: Techniques of Vacuum Ultraviolet Spectroscopy. Wiley, New York 1967.

[16] D.A. Huchital and J.D. Rigden, Appl. Phys. Lett. *16*, 348 (1970).

[17] a) P.H. Citrin, P.W. Shaw and T.D. Thomas in [13], p. 105.
     b) K. Maeda, ibid. p. 177.

[18] B. Feuerbacher and B. Fitton, Phys. Rev. Lett. *29*, 786 (1972).

[19] B.J. Waclawski and E.W. Plummer, Phys. Rev. Lett. *29*, 783 (1972).

[20] D.E. Eastman and W.D. Grobman, Phys. Rev. Lett. *28*, 1378 (1972).

[21] K. Siegbahn in [13], p. 15.

[22] G.K. Wertheim and S. Hüfner, Phys. Rev. Lett. *28*, 1028 (1972).

[23] C.N. Berglund and W.E. Spicer, Phys. Rev. *136*, 1030 (1964).

[24] N.V. Smith, Phys. Rev. *B3*, 1862 (1971).

[25] D.E. Eastman in [13], p. 487.

[26] Y. Baer, F. P. Heden, J. Hedman, M. Klasson, C. Nordling and K. Siegbahn, Solid State Comm. *8*, 517 (1970).

[27] D. A. Shirley in [13], p. 603.

[28] S. B. M. Hagström in [13], p. 515.

[29] R. A. Pollack, S. Kowalczyk, L. Ley and D. A. Shirley, Phys. Rev. Lett. *29*, 274 (1972).

[30] A. R. Williams, J. F. Janak and V. L. Moruzzi, Phys. Rev. Lett. *28*, 671 (1972).

[31] D. T. Pierce and W. E. Spicer, Phys. Rev. Lett. *25*, 581 (1970).

[32] J. E. Rowe and J. C. Tracy, Phys. Rev. Lett. *27*, 799 (1971).

[33] a) P. Soven, Phys. Rev. *178*, 1136 (1969).
   b) B. Velický, S. Kirkpatrick and H. Ehrenreich, Phys. Rev. *175*, 747 (1968).

[34] a) S. Hüfner, G. K. Wertheim, R. L. Cohen and J. H. Wernick, Phys. Rev. Lett. *28*, 488 (1972).
   b) S. Hüfner, G. K. Wertheim, J. H. Wernick an A. Melera, Solid State Comm. *11*, 259 (1972).

[35] S. Hüfner, G. K. Wertheim and J. H. Wernick, Phys. Rev. *B8*, 4511 (1973).

[36] J. Friedel, Nuovo Cimento (suppl.) *7*, 281 (1958).

[37] P. W. Anderson, Phys. Rev. *124*, 41 (1961).

[38] a) U. Gerhardt and E. Dietz, Phys. Rev. Lett. *26*, 1477 (1971).
   b) R. Y. Koyama and L. R. Hughey, ibid. *29*, 1518 (1972).

[39] a) U. Bänninger, G. Busch, M. Campagna and H. C. Siegmann, Phys. Rev. Lett. *25*, 585 (1970).
   b) G. Busch, M. Campagna, D. T. Pierce and H. C. Siegmann, ibid. *28*, 611 (1972).

[40] a) W. L. Schaich and N. W. Ashcroft, Phys. Rev. *B3*, 2452 (1971).
   b) G. D. Mahan, Phys. Rev. *B2*, 4334 (1970).

[41] R. Haydock, V. Heine, M. J. Kelly and J. B. Pendry, Phys. Rev. Lett. *29*, 868 (1972).

[42] H. D. Hagstrum and G. E. Becker, Phys. Rev. *159*, 572 (1967).

[43] F. Cyrot-Lackmann, J. Phys. Chem. Solids *29*, 1235 (1968).

[44] a) F. Forstmann and V. Heine, Phys. Rev. Lett. *24*, 1419 (1970).
   b) F. Forstmann and J. B. Pendry, Z. Phys. *235*, 75 (1970).

[45] D. E. Eastman, Phys. Rev. *B3*, 1769 (1971).

[46] L. Ley, S. Kowalczyk, R. A. Pollack and D. A. Shirley, Phys. Rev. Lett. *29*, 1088 (1972).

[47] a) R. A. Pollack, L. Ley, S. Kowalczyk, D. A. Shirley, J. D. Joannopoulos, D. J. Chudi and M. L. Cohen, Phys. Rev. Lett. *29*, 1103 (1972).

[48] E. W. Plummer and R. D. Young, Phys. Rev. *B1*, 2088 (1970).

[49] a) E. W. Plummer and J. W. Gadzuk, Phys. Rev. Lett. *25*, 1493 (1970).
   b) J. W. Gadzuk, J. Vac. Sci. Techn. *9*, 590 (1972).

[50] B. J. Waclawski, L. R. Hughes and R. P. Madden, Appl. Phys. Lett. *10*, 305 (1967).

[51] W. T. Bordass and J. W. Linnett, Nature *222*, 660 (1969).

[52] D. R. Penn, Phys. Rev. Lett. *28*, 1041 (1972).

[53] D. E. Eastman, Solid State Comm. *10*, 933 (1972).

[54] D. E. Eastman, J. K. Cashion and A. C. Switendick, Phys. Rev. Lett. *27*, 35 (1971).

[55] A. K. Kress and G. J. Lapeyre, Phys. Rev. Lett. *28*, 1639 (1972).

[56] C. R. Helms and W. E. Spicer, Phys. Rev. Lett. *28*, 565 (1972).

[57] W. D. Grobmann and D. E. Eastman, Phys. Rev. Lett. *29*, 1508 (1972).

[58] a) A. Many, Y. Goldstein and N. B. Grover: Semiconductor Surfaces. North Holland, Amsterdam 1965.
   b) D. R. Frankel: Electrical Properties of Semiconductor Surfaces. Pergamon Press, Oxford 1967.
   c) W. Mönch, in: Festkörperprobleme Vol. 13 (H. J. Queisser, ed.), Vieweg, Braunschweig 1973.

[59] L. F. Wagner and W. E. Spicer, Phys. Rev. Lett. *28*, 1381 (1972).

[60] D. E. Eastman and W. D. Grobman, Phys. Rev. Lett. *28*, 1378 (1972).

[61] D. L. Smith and D. A. Huchital, J. Appl. Phys. *43*, 2624 (1972).

[62] C. R. Brundle and M. W. Roberts, Proc. Roy. Soc. (London) *A331*, 383 (1972).

[63] see for example:
   a) C. D. Wagner and P. Biloen, Surface Sci *35*, 82 (1973).
   b) G. Schön, Surface Sci. *35*, 96 (1973).

[64] Commercially available by McPherson.

[65] T. E. Madey, J. T. Yates and N. E. Erickson, Chem. Phys. Lett. *19*, 487 (1973).

5. Appearance Potential Spectroscopy (APS)

5.1. Introduction

The methods which were described in the preceding chapters are based on the emission of electrons caused by primary excitation by means of electrons and photons. It is also possible to use low energy electrons with their low penetration depth into solids to create photons in the surface region. In this case the energy of the primary electrons which create soft x-rays has to be restricted to the range of 10–1000 eV.

Soft x-ray spectroscopy has found considerable interest in solid state physics in the past as a tool for determining electronic band structures [1]. However the energy of the primary electrons exceeds several keV in most of these experiments, so that the information originates mainly from the bulk. Since the dispersion of x-rays using a grating or crystal spectrometer presents considerable experimental problems if it is to be made compatible with UHV techniques, very little attempt has been made to study surface properties by analyzing soft x-ray spectra [3]. It is however possible to use such a technique for surface analysis, provided the primary energy of excitation is lower than about 2000 eV. This has been done recently by Sewell et al. [37] who studied the initial stages of oxidation of a Ta(110) surface using oxygen K_α x-ray emission. Even the emission of visible light from electron-irradiated surfaces might probably be used for a surface spectroscopy.

An important variation of soft x-ray emission spectroscopy is the so-called "isochromat spectroscopy" whereby the change in intensity is recorded for one fixed wavelength as a function of the electron excitation energy [2]. Such spectra yield information about the unoccupied electron states above the Fermi level.

Similar information may be obtained by appearance potential spectroscopy (APS) in a much simpler manner:

By measuring the total x-ray emission intensity (without dispersion) as a function of the primary electron energy the threshold energies for the excitation of core electrons to the Fermi level may be determined from the "appearance" of small bumps in the intensity vs. primary energy curve when the primary energy corresponds to the energy of a particular characteristic x-ray. This technique was used extensively in the 1920's to determine the energies of core states in the range between 50 and 1000 eV [4]. The applicability of APS for surface studies was demonstrated by Park, Houston and Schreiner [5] in 1970 using an ingeniously simple experimental device [6]. Redhead [9] had pointed out the high sensitivity of APS to the state of the surface proposing the term "non-dispersive spectroscopy" some years previously. Since Auger-electron emission and x-ray fluorescence compete for the de-excitation of core holes it is possible to measure either the total x-ray intensity (SXAPS) or the total Auger-electron yield (AEAPS) [7]. The latter is essentially just the derivative of the secondary emission coefficient with respect to incident electron energy, and has certain disadvantages compared to SXAPS. The following is concerned mainly with SXAPS (APS for short). Since APS probes the empty states above the Fermi level this method is in some sense complementary to photoelectron spectroscopy. It has been applied for studies of the core level energies of surface atoms, the electronic structure of metals and alloys, and plasmon coupling, as well as for the investigation of chemical effects on the electronic states of surface atoms. The applicability as an analytical tool for determining the surface composition in a manner similar to AES is somewhat limited, since APS is quite insensitive to a number of elements.

A critical survey of some of the aspects of APS has been published by Tracy [8], which, however, leaves some points open to discussion [38]. More recently two review papers on APS were written [39, 40].

5.2. Experimental

A typical APS spectrometer for surface studies is quite simple and is shown schematically in fig. 5.1. Electrons are emitted from the cathode (usually a tungsten filament) and are accelerated

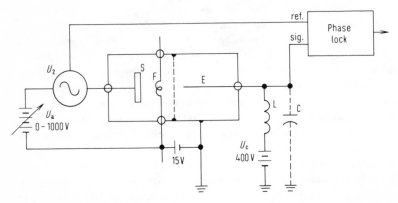

Fig. 5.1. Scheme of a spectrometer for SXAPS. S: sample, F: filament, E: collecting electrode, L: inductor with high Q, C: distributed capacitance. After Park and Houston[6].

by the variable voltage U_a onto the sample. The resulting x-rays pass through a grid and enter the collector, whereas both secondary electrons originating from the target and primary electrons from the filament are repelled by the grid, to which is applied a negative potential. The x-radiation creates photoelectrons on the walls which serve as photocathode. The electrons are then picked up on the collector E which is at some positive potential U_c.

In order to improve the sensitivity the electronic modulation technique is again applied. An ac voltage U_2 (amplitude a few tenths of a volt, frequency ~ 1 to 10 kHz) is superimposed on the slowly varying acceleration voltage U_a and the signal with frequency ω is selected and amplified with a lock-in amplifier.

The nondifferentiated intensity vs. voltage curve exhibits a continuous increase due to bremsstrahlung with very small peaks due to the onset of characteristic radiation. Electronic differentiation is therefore used exactly as with AES. If the objective is simply to determine a chemical analysis of the surface, it may be advantageous to degrade the instrumental resolution by increasing the amplitude of the potential modulation and detect the second harmonic of the modulating frequency.

With primary electron currents between 1 to 10 mA photocurrents in the range of 10^{-11} to 10^{-12} A are obtained. Suitable screening and insulation must therefore be provided for the collector.

Musket and Taatjes [10] described a compact and simple APS spectrometer which is shown in fig. 5.2. The stainless steel walls of the analyzer serve as photocathode as proposed by

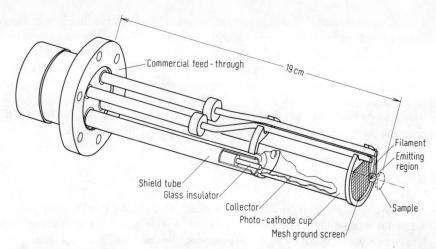

Fig. 5.2. Cut-away view of the compact appearance-potential spectrometer described by Musket and Taatjes [10].

Houston and Park [6]. Haas et al. [11] used a conventional LEED optics as an analyzer for APS, in which the x-rays emitted from the sample eject electrons from the fluorescent screen. Another construction was proposed by Long and Beavis [12]. Their detector circuit is near ground potential which permits operation under pressures greater than 10^{-7} Torr, owing to the absence of stray currents caused by positive ions.

The limit of detection is given by the shot noise of the photocurrent produced by the bremsstrahlung background. Since the signal is proportional to the collector current i_c, but the shot noise proportional only to $\sqrt{i_c}$, the signal to noise ratio can be improved by increasing i_c. This can be done simply by increasing the primary electron current, which is limited by the maximum temperature rise of the sample surface which can be tolerated. This effect may be a considerable nuisance in studies with adsorbate covered surface. Another possibility is to use other materials for the photocathodes with higher photo yields than stainless steel [13]. Some possible artefacts in APS spectra caused by the influence of the photocathode material were discussed by Tracy [8].

A device for Auger electron-APS is shown schematically in fig. 5.3. Electrons ejected from the sample move in a field-free region and those emerging at nearly grazing angles pass through

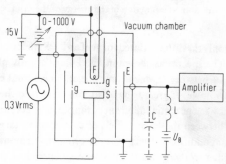

Fig. 5.3. Scheme of a spectrometer for AEAPS. S: sample, F: filament, E: electrode for collecting secondary electrons, g: grids for maintaining a field free region around the sample. The bias U_B is reversed for SXAPS. After Houston and Park [7].

a slit in the shield surrounding the sample and are collected on the electrode E. The collector is biased by $U_B \approx 200$ V in order to reject the low energy "true" secondary electrons. A small ac voltage is superimposed on the potential between cathode and sample in order to obtain the derivative of the collector current. A high-Q inductor L in parallel with the capacitance C forms a resonant circuit at the ac frequency. Since the signals are about 10^4 larger than those observed in SXAPS no lock-in amplifier is needed. In contrast to SXAPS the background with AEAPS is not well behaved. The same device may also be used for SXAPS simply by reversing the bias U_B on the collector which then acts as a photocathode.

5.3. The mechanism of APS

The total intensity of x-rays emitted from an electron bombarded sample consists of a smoothly increasing bremsstrahlung background on which is superimposed the characteristic radiation. The latter is proportional to the rate of excitation of an electron from a core level to an unfilled state above the Fermi level. An example is shown in fig. 5.4, from which it is evident that

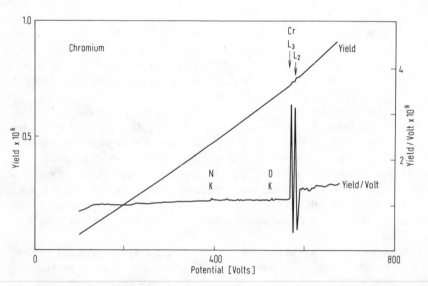

Fig. 5.4. Comparison of the total x-ray yield and the differential yield (=APS spectrum) of chromium as a function of the energy of the primary electrons. After Houston and Park [22].

only the differentiated form of an APS spectrum is sensitive enough to yield detailed information. The spectra as shown in fig. 5.4 are usually plotted as a function of the accelerating potential U_a between filament and target. The actual energy E_p of an electron incident on the sample is however greater by an amount equal to the work function $e\phi_c$ of the filament as illustrated by the potential diagram of fig. 5.5. The accelerating voltage determines the potential difference between the Fermi levels of cathode and target, but electrons leaving the filament must have an additional energy $e\phi_c$ in order to surmount the work function barrier. E_p is spread by ΔE due to the mean thermal energy spread of the emitted electrons. ΔE is usually in the range

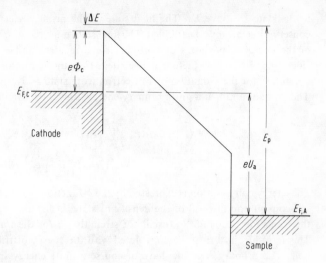

Fig. 5.5. Potential diagram for APS experiments.

of ~ 0.5 eV. Assuming that only one-electron processes are involved (the possible participation of collective electron excitations is demonstrated below), the mechanism is illustrated in fig. 5.6, where E_B is the energy of the core level, $N_c(E' - E_B)$ is the density of states around E_B, which is mainly determined by the lifetime of the core-hole, and $N(E)$ is the density of the valence band states, which are unfilled above the Fermi energy E_F. E_F is taken as the energy zero. The incoming electron with primary energy E_p loses energy to the core hole which is excited

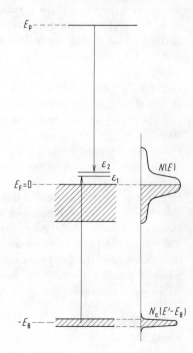

Fig. 5.6. Energy level diagram for APS.

to the state ε_1 above E_F. The final state of the primary electron is ε_2, whereby the energy conservation law must be fulfilled. The final state must therefore take all possible combinations of the energetic positions of both electrons (incident and excited core electron) into consideration. From this picture it is evident that the minimum primary energy ('appearance potential') for the excitation of an electron from state E_B is simply given by $E_p = E_B$.

The excitation rate at a given primary energy E_p will be given by [14]

$$P(E_p) \propto \int_0^{E_p} N_c(E' - E_B)\,dE' \int_0^{E_p + E'} P(E' \to \varepsilon_1) \cdot P(E_p \to \varepsilon_2) \cdot$$

$$\cdot N(\varepsilon_1) N(\varepsilon_2)\, \delta\, (\varepsilon_1 + \varepsilon_2 - E' - E_p)\,d\varepsilon_1$$

(5.1)

The energy conservation is expressed by the δ-function, which turns the second integral into a self-convolution integral of the density of states $N(E)$ above the Fermi level, multiplied by the transition probabilities (oscillator strengths) P for the transitions $E' \to \varepsilon_1$ and $E_p \to \varepsilon_2$. This term has in turn to be convoluted with the energy distribution $N_c(E' - E_B)$ of the core hole. The primary electrons have of course a finite energy spread ΔE around E_p which in principle requires a further convolution.

If we assume that ΔE and the energy width of the core hole may be neglected and that the oscillator strengths are constant (i.e. they do not vary considerably with variation of ε_1 across the allowed portion of $N(E)$ above E_F) then equ. (5.1) simplifies to

$$P(E_p) \propto N_c(E_B) \cdot \int_{0(=E_F)}^{E} N(\varepsilon_1)\, N(E - \varepsilon_1)\,d\varepsilon_1$$

(5.2)

The initial energy $E_p = |E_B| + E$ exceeds the binding energy of the core state E_B by the energy E. If $E_p > |E_B|$ the primary electron can transfer an amount of energy $E_B + \varepsilon_1$ to the core electron which is then excited to the state ε_1 above E_F, the primary electron retains with $\varepsilon_2 = E_p - (|E_B| + \varepsilon_1) = E - \varepsilon_1$. The upper limit of the integral corresponds to the situation where the core electron is just excited to the Fermi level; the remaining primary electron energy is $E_p - E_B$. The lower limit is given by the reverse situation, i.e. the primary electron at E_F and the excited core electron at $E_p - E_B$. The integral expresses all possible combinations and represents therefore the density of the final states, whereas $N_c(E_B)$ is the density of the initial state.

Finite values for the integrand in equ. (5.2) can only be obtained if the density of states for the energy ε_1 as well as for $\varepsilon_2 = E - \varepsilon_1$ is large enough. It is thus evident that the intensity of the emitted x-radiation is sensitive to the density of unfilled states above the Fermi level. It is further important to notice that this method probes a local density of states in the vicinity of those nuclei whose core electrons are excited. The degree of localization depends on the extension of the particular core level wave function.

The second and possibly even more important factor determining the intensities concerns the fluorescent yield for the de-excitation process. This quantity may vary considerably thus giving very different sensitivities to different transitions.

APS spectra contain (at least in principle) the following information:

a) The binding energy of core levels may be determined from the corresponding appearance potential.

b) The density of states at the Fermi level is correlated with the height of the differentiated APS peaks.

c) The shape of the APS peaks is related to the electronic structure of surface atoms and may vary with their chemical state.

5.4. Surface analysis

Since the appearance potentials represent just the binding energies of the corresponding core electrons APS seems to be a very simple and powerful method for surface analysis. However direct comparisons [15], [16] between APS and AES revealed that the latter is much more sensitive for most elements and that APS is virtually unable to detect some elements.

APS has a particularly low sensitivity to the common surface impurities like S, O and C; it is sensitive to all 3 d metals with unfilled d-band, but very insensitive to metals with filled d-bands like Cu. APS is almost unable to detect 4 d- and 5 d-transition metals with unfilled d-states like Pd [8], but the sensitivity for a 6 d element like Th is very high [17] as it is for the rare-earth elements [18]. This applies to excitations from states with energies below ~1000 eV since these are the transitions which are of interest in surface analysis. For example it is rather difficult to study the N levels from 5 d elements whereas their M levels produce intense spectra. A further disadvantage of APS as compared with AES is that primary electron currents > 1 mA are necessary which may heat the sample surface to temperatures above 200 °C and interact strongly with adsorbed layers. It would therefore seem to be almost impossible to study adsorbed phases by means of APS.

A possible analytical application of this method had been pointed out by Park et al. [19] and concerns the surface composition of steels in connection with problems of corrosion. Using samples of Cr, Fe, Ni and 304 stainless steel under identical conditions the authors established a reasonable calibration simply by comparing the peak heights. The results revealed that annealing at higher temperature leads to a depletion of chromium on the surface, an effect which might be responsible for the behaviour of this material towards corrosion.

5.5. Energies of atomic core levels

The binding energies of the core electrons with respect to the Fermi energy follow directly from measurements of the 'appearance potentials', i.e. the excitation edges for APS peaks with addition of the work function of the filament (a further small correction of about 0.5 eV takes account of the thermal energy spread of the primary electrons and of the amplitude of the modulation voltage). A systematic investigation of the binding energies of L_3-electrons of the 3 d-transition metals has been made by Park and Houston [20]. Their results are given in table 5.1 together with tabulated values by Bearden and Burr [21]. It is important to notice that the binding energies so determined are always somewhat lower than the literature values. This behaviour had already been observed by Dev and Brinkman [14] using a somewhat different experimental technique and who ascribed this effect to the fact that the reduced coordination of the surface atoms should produce a "chemical shift" in their core electrons to lower binding energies.

Park and Houston [20] also determined the spin-orbit splittings between L_3 and L_2 levels. Their data are in excellent agreement with tabulated x-ray values [21].

Table 5.1. Binding energies of the L_3-electrons of the 3d-metals. Comparison of APS results [20] with tabulated x-ray values [21]. After Park and Houston [20].

	L_3 (2p$_{3/2}$)		L_2 (2p$_{1/2}$)		L_1 (2s)	
	Park & Houston [20]	Bearden & Burr [21]	Park & Houston [20]	Bearden & Burr [21]	Park & Houston [20]	Bearden & Burr [21]
21 Sc	398.4±0.5	402.2±0.4	402.1±0.5	406.7±0.4	496.4±0.7	500.4±0.4
22 Ti	453.4±0.5	455.5±0.4	459.3±0.5	461.5±0.4	558.4±0.6	563.7±0.4
23 V	512.6±0.5	512.9±0.3	519.0±0.5	520.5±0.3	625.8±0.7	628.2±0.4
24 Cr	574.0±0.5	574.5±0.3	582.7±0.5	583.7±0.3	694.1±0.7	694.6±0.4
25 Mn	638.5±0.5	640.3±0.4	649.1±0.5	651.4±0.4	766.2±0.7	769.0±0.4
26 Fe	706.3±0.5	708.1±0.9	719.3±0.5	721.1±0.9	842.7±0.7	846.1±0.4
27 Co	776.9±0.5	778.6±0.3	792.0±0.5	793.6±0.3	992.6±0.7	925.6±0.4
28 Ni	850.9±0.5	854.7±0.4	868.2±0.5	871.9±0.4	1007.4±0.7	1008.1±0.4

The width of the APS peaks is determined by the structure of the unfilled portion of the valence band and by the energy spread of the core hole. The latter depends on the lifetime of the core hole due to the action of the uncertainty principle. Since the fluorescent yield is very small for $|E_B| < 1\,000$ eV the lifetime is determined by the Auger rate for that core hole. In particular, strong Coster-Kronig transitions from $2p \rightarrow 2s$ states occur, giving rise to very short lifetimes of holes in the L_1 state. This effect becomes evident from a comparison of the Cr L_3 and L_1 spectra in fig. 5.7 [20]. Whereas the L_3 width is almost entirely due to the width of the unfilled d-band, the L_1 peak is much broader due to the short lifetime of the core hole.

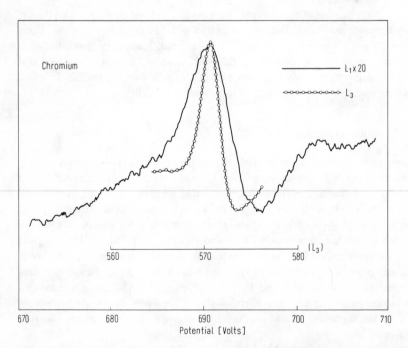

Fig. 5.7. Comparison of the shapes of the L_1 and L_3 peaks of chromium. After Park and Houston [20].

5.6. Band structure of metals and alloys

Information about the structure and density of unfilled states above the Fermi level is contained in the shapes and heights of the APS peaks. Assuming constant oscillator strength and negligible core hole lifetime broadening one arrives at equ. (5.2) which describes the dependence of the total intensity of characteristic x-radiation on a smooth background of bremsstrahlung as a function of the energy E_p of the primary electrons. If the actual band structure of a transition metal with unfilled d-band is approximated schematically by a rectangular shape for the densities of empty d-band and sp states as shown in fig. 5.8a [22] then the convolution operation

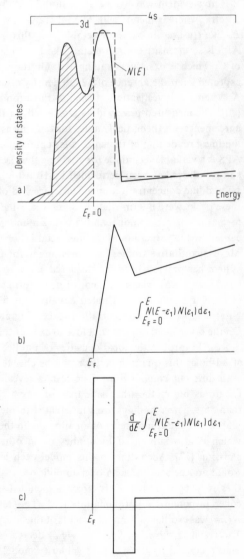

Fig. 5.8. Simplified scheme for the formation of an APS peak (without lifetime broadening of the core hole).
a) density of states $N(E)$ of the valence band
b) convolution of $N(E)$ according to equ. (5.2)
c) resulting appearance potential spectrum.
After Houston and Park [22].

of equ. (5.2) can be carried out easily. The result is shown in fig. 5.8b which represents the shape of the total x-ray emission curve. In APS the spectra are measured as the derivative with respect to the accelerating voltage and therefore the shape of the APS spectrum of a metal with unfilled d-band is expected to look like that in fig. 5.8c: a peak followed by a somewhat smaller undershoot and a step for each of the core levels. The width of the positive peak should be approximately equal to the width of the unfilled d-band, provided that other broadening effects like core-hole lifetime and primary electron spread do not dominate. In the absence of the s-band, the negative peak would be equal to the positive, so that the reduction of the negative dip is thus a measure of the relative contribution of the s-states. The height of the positive peak is proportional to $N^2(E_F)$, where $N(E_F)$ is the density of states at the Fermi level. It is now evident why 3d metals with unfilled d-states like Cr or Ni exhibit pronounced APS spectra, whereas metals with filled d-states and only low densities of states of s-electrons at E_F like Cu are almost undetectable with this method, but the model does not explain why APS is so insensitive to 4d metals like Pd. Fig. 5.7 shows the L_3 peak of the APS spectrum of Cr which is exactly as anticipated. On the other hand Cu reveals only a small step [8] as expected from the absence of a high density of unfilled d-states at the Fermi level.

A systematic investigation of the 3d-transition metals has been made by Park and Houston [20]. The measured peak widths (corrected for the instrument broadening) are in reasonable agreement with theoretical values for the widths of the unfilled d-bands. With Fe, Co and Ni lifetime broadening of the core holes contributes considerably to the observed peak widths. APS is a valuable technique for studying the electronic structure of alloys. A study with Ti-Ni surface alloys has been performed by Houston and Park [23] who found identical peaks of Ni in dilute concentration on a Ti surface and of dilute coverages of Ti on Ni. The widths of these peaks were intermediate between the values for clean Ni and Ti. The results suggest the formation of a common band, i.e. a site-independent density of states which is probably caused by the formation of an intermetallic compound TiNi.

More quantitative studies have been made for Cu/Ni alloys by Ertl and Wandelt [24]. This system has previously been considered as being a good example for the validity of the rigid-band model [25], where both constituents form common bands and the additional s-electrons of the Cu atoms are distributed equally thus filling continuously the empty d-states of the Ni atoms. With Cu concentrations exceeding about 55 at.% all Ni d-states should be filled. On the other hand d-holes should also exist at the sites of Cu atoms at low Cu concentrations. Since APS probes the local densities of unfilled states and is very sensitive to the existence of 3d-holes this question could easily be checked by recording APS spectra of Cu/Ni alloys with different compositions. The results revealed that in no case d-holes enter the sites of Cu atoms, but on the other hand that d-holes at Ni sites exist even at Cu concentrations greater than 55%. These findings are consistent with the predictions of the minimum polarity model [26], where the components of an alloy retain the electronic configurations of the pure phases which is also supported by studies using other experimental methods, e.g. x-ray photoemission [27]. According to this model each lattice site remains electrically neutral which would appear also to be an oversimplification, since a partial charge-transfer from Cu to Ni sites is to be expected. This effect is included in more refined theories on the basis of the coherent-potential-approximation [28]. The relative density of states at the Fermi level $N(E_F)$ was evaluated from the heights of the Ni L_3 peaks as a function of the alloy composition. The result is shown in fig. 5.9. $N(E_F)$ decreases more strongly with increasing Cu content than predicted by the minimum polarity model (MP) but the data agree very well with the

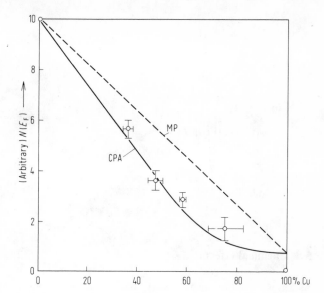

Fig. 5.9. Density of states at the Fermi level $N(E_F)$ for a series of Cu/Ni alloys as determined by APS as compared with theoretical predictions of the minimum polarity model (broken line) and of calculations on the basis of the 'coherent potential approximation' (full line). After [24].

theoretical curve calculated by Stocks et al. [29] on the basis of the coherent-potential-approximation. This problem could not be solved unequivocally by photo-electron spectroscopy [27], since there the steep decrease of the density of states at E_F causes a high degree of uncertainty.

5.7. Plasmon coupling

So far the processes involved in APS have been treated on the basis of the single-electron picture as outlined in section 5.3. The results with 3d metals reveal that the assumption is justified to a good approximation for this class of elements. It has been suggested however by Langreth [30] that in APS where the final state before deexcitation includes a core hole and two electrons near the Fermi level, plasmon coupling may play an important role, which means that coupling between the suddenly created core hole to the collective excitation of conduction electrons (plasmon) produces structure in the spectra which is displaced from the principal peak by the energy needed to excite the plasmon. This effect has been examined theoretically by Laramore [31].

The appearance of these satellite peaks has been observed experimentally with light elements like carbon [32, 33], boron [32], and oxygen [36]. The appearance potential spectrum corresponding to K-shell excitation of carbon [33] is shown in fig. 5.10. One can clearly distinguish six sharp peaks which are separated from one another by about 7 eV. This value corresponds to the plasmon energy in graphite. The first peak is therefore interpreted as due to excitation to a state where the collective modes for the conduction electrons remain in the ground state, the second peak indicates the emission of a single plasmon and so on. More detailed investigations revealed that the energetic distances between the different peaks vary slightly [35]. Fine structure is sometimes also observed in the APS spectra of other elements, but definite association with plasmon excitations is usually not possible since characteristic energy losses due to interband transitions must also be taken into consideration.

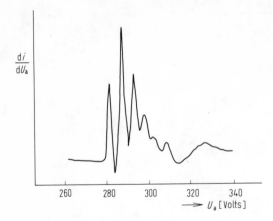

Fig. 5.10. Appearance potential spectrum of carbon with satellite peaks owing to plasmon excitations. After Houston and Park [33].

5.8. Chemical effects

As known from extensive ESCA studies the formation of chemical bonds may change the binding energies of core electrons. Similar effects are to be expected with APS leading to "chemical shifts" of the threshold energies. But in contrast to ESCA the excited states with APS are in the valence band which is of course also strongly affected by the chemical bond.

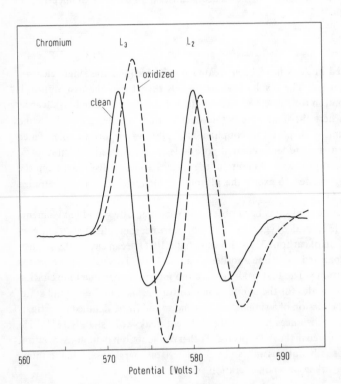

Fig. 5.11. Effect of oxidation on the L_3/L_2 spectrum of chromium. After Houston and Park [36].

It is therefore not possible to compare directly variations in the core level binding energies with respect to the Fermi level as determined by APS with traditional 'chemical shifts' (with respect to the vacuum level) as derived from ESCA.

Variations of the threshold energies for the APS peaks of some 3d metals have been observed after interaction with oxygen [22]. The effect of oxygen in each case consists in a shift of the binding energies to about 0.5–2.0 eV higher values. (Although the determination of absolute thresholds is accurate only to about 0.5 eV, variations in the order of 0.1 eV are easily detectable). APS spectra of the L_3 and L_2 peaks of clean and oxidized chromium are shown in fig. 5.11 [36]. Besides the shift of the threshold energy a variation in the shape of the peaks is clearly detectable. With the oxidized samples the negative dips are larger and the peak widths (and therefore the apparent width of the unoccupied d-band) increase by about 0.8 eV. This effect is interpreted as a partial removal of electrons from Cr atoms by oxygen.

5.9. References

[1] D. Fabian, in: CRC Critical Reviews in Solid State Science, Aug. 1971, p. 255.
[2] H. Merz and K. Ulmer, Z. Phys. *212*, 435 (1968).
[3] P.B. Sewell and M. Cohen, Appl. Phys. Lett. *11*, 298 (1967).
[4] H.W.B. Skinner, Proc. Roy. Soc. *135 A*, 84 (1932).
[5] R.L. Park, J.E. Houston and D.G. Schreiner, Rev. Sci. Instr. *41*, 1810 (1970).
[6] R.L. Park and J.E. Houston, Surface Sci. *26*, 664 (1971).
[7] J.E. Houston and R.L. Park, Phys. Rev. *B5*, 3808 (1972).
[8] J.C. Tracy, J. Appl. Phys. *43*, 4164 (1972).
[9] P.A. Redhead, Bull. Radio and Elec. Eng. Dir. Natl. Res. Council Canada, *16*, 41 (1966).
[10] R.G. Musket and S.W. Taatjes, J. Vac. Sci. Techn. *9*, 1041 (1972).
[11] T.W. Haas, S. Thomas and G.J. Dooley, Surface Sci. *28*, 645 (1971).
[12] R.L. Long and L.C. Beavis, Rev. Sci. Instr. *43*, 939 (1972).
[13] A.H. Sommer: Photoemissive Materials. John Wiley, New York 1968.
[14] B. Dev and H. Brinkman, Nederl. Tijdschr. Vacuum-techniek *8*, 176 (1970).
[15] J.C. Tracy, Appl. Phys. Lett. *19*, 353 (1971).
[16] R.G. Musket, J. Vac. Sci. Techn. *9*, 603 (1972).
[17] P.A. Redhead and G.W. Richardson, J. Appl. Phys. *43*, 2970 (1972).
[18] R.L. Park and J.E. Houston, J. Vac. Sci. Techn. *10*, 176 (1973).
[19] R.L. Park, J.E. Houston and D.G. Schreiner, J. Vac. Sci. Techn. *9*, 1023 (1972).
[20] R.L. Park and J.E. Houston, Phys. Rev. *B6*, 1073 (1972).
[21] J.A. Bearden and A.F. Burr: Atomic Energy Levels. US Atomic Energy Commission, NYO 2543-1, Oak Ridge 1965.
[22] J.E. Houston and R.L. Park, J. Chem. Phys. *55*, 4601 (1971).
[23] J.E. Houston and R.L. Park, J. Vac. Sci. Techn. *9*, 579 (1972).
[24] G. Ertl and K. Wandelt, Phys. Rev. Lett. *29*, 218 (1972).
[25] N.F. Mott, Proc. Phys. Soc. London *47*, 571 (1935).
[26] N.D. Lang and H. Ehrenreich, Phys. Rev. *168*, 605 (1968).
[27] S. Hüfner, G.K. Wertheim, R.L. Cohen and J.H. Wernick, Phys. Rev. Lett. *28*, 488 (1972).
[28] S. Kirkpatrick, B. Veličky and H. Ehrenreich, Phys. Rev. *B1*, 3250 (1970).
[29] G.M. Stocks, R.W. Williams and J.S. Faulkner, Phys. Rev. *B4*, 4390 (1971).
[30] D.C. Langreth, Phys. Rev. Lett. *26*, 1229 (1971).
[31] G.E. Laramore, Solid State Comm. *10*, 85 (1972).
[32] J.E. Houston and R.L. Park, J. Vac. Sci. Techn. *8*, 91 (1971).
[33] J.E. Houston and R.L. Park, Solid State Comm. *10*, 91 (1972).
[34] D.L. Greenaway, G. Harbeke, F. Bassani and E. Tasatti, Phys. Rev. *178*, 1340 (1969).
[35] A.M. Bradshaw and D. Menzel, phys. stat. sol. (b) *56*, 135 (1973).

[36] J. E. Houston and R. L. Park, in: Electron Spectroscopy. (D. A. Shirley, ed.) North Holland, Amsterdam 1972.

[37] P. B. Sewell, D. F. Mitchell and M. Cohen, Surface Sci. *29*, 173 (1972).

[38] R. L. Park and J. E. Houston, J. Appl. Phys. *44*, 3810 (1973).

[39] R. L. Park and J. E. Houston, J. Vac. Sci. Techn. *11*, 1 (1974).

[40] A. M. Bradshaw, in: Surface and Defect Properties of Solids, Chemical Society, London 1974.

6. Field Emission Spectroscopy

6.1. Introduction

Tunneling of electrons through the surface barrier under the influence of an external electric field forms the basis of the field emission and field ion microscopes invented by E. W. Müller [1, 2]. Although strictly these techniques both fall within the scope of this work they are not treated here. The reader is referred to a series of excellent monographs already published on this topic [2–4].

Besides imaging of the surface, field electron emission can also be used to determine the work function of the emitting area (as described in 8.2.3) and to provide some information about the density of electronic states in the surface region. This latter information is derived from measurements of the energy distribution of field emitted electrons. Electric field strengths of the order of 10^7 V/cm are necessary for appreciable field emission, and this technique is therefore restricted to surfaces with very small radii of curvature, e. g. fine wires etched to sharp tips with cross sections of only a few hundred Ångstroms, as such fields are to all intents and purposes unobtainable at more conventional surfaces.

The energy distribution of field emitted electrons is usually measured by the retarding field technique. It is important with this method that the electrons strike the collector of the analyzer at normal incidence, since otherwise the energy resolution is degraded. Fig. 6.1 shows a

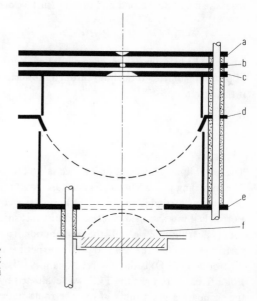

Fig. 6.1. Field electron energy analyzer
a: anode, b: lens electrode, c: shielding electrode, d: hemispherical grid, e: output grid, f: first dynode of electron multiplier. After Czyzewski [15].

device as developed by Czyzewski [15] similar to that by van Oostrom [5] consisting of a probe hole which transmits the electrons emitted from a particular region of the surface. These electrons are then focused by an electrostatic lens into a hemispherical Faraday cage.

An extended review on the subject of field emission spectroscopy was written by Gadzuk and Plummer [25].

6.2. Energy distribution of field emitted electrons from clean surfaces

The basic features of the field emission process may be described by a one-dimensional model of the electrostatic potential near the surface as illustrated in fig. 6.2. In the presence of a high

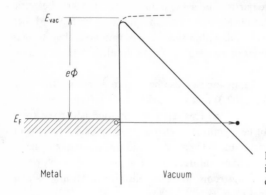

Fig. 6.2. Potential diagram at the metal/vacuum interface in the presence of a strong external electric field.

external electric field the potential at the surface is deformed in such a way that a barrier of finite height and finite thickness arises which the metallic electrons may penetrate via the tunnel effect. The number of electrons $\dot{n}(E)\,dE$ with energy between E and $E+dE$ escaping from the metal per second and per cm^2 surface area is given by [3, 6]

$$\dot{n}(E)\,dE = N(E) \cdot P(E)\,dE \tag{6.1}$$

where $N(E)\,dE$ is the energy distribution of electrons inside the solid and $P(E)$ the probability of transmission (i.e. tunneling) through the surface barrier. The total emitted current density is then given by

$$i = e \int_{E_1}^{\infty} \dot{n}(E)\,dE \tag{6.2}$$

The lower limit of integration E_1 is equal to the lowest electronic state in the solid.

The tunneling probability $P(E)$ decreases very rapidly with decreasing energy, so that in fact only those states which are near the Fermi level contribute significantly to the emitted current. Formerly it was assumed that the quantity measured by a retarding-field device was the normal energy distribution, i.e. only the contribution from electrons with a momentum normal to the surface. But in fact the geometry of the experimental arrangement is such that the total energy distribution (TED) is the observed quantity [7].

For a free electron gas the TED was evaluated by Young [7] using Fermi-Dirac statistics for the occupation probability of the electronic states. With the origin of the energy scale taken at the Fermi level the result is

$$i(E) = i_0 \cdot \exp\,(E/\beta)/[1 + \exp\,(E/kT)] \tag{6.3}$$

in which i_0 is the total current density of field emitted electrons and depends on the field strength F and the work function ϕ as given by the Fowler-Nordheim equation (8.11); $\beta =$

$=\hbar e F/2(2m\phi)^{1/2} t(y)$ where $t(y)$ is a tabulated correction function [6]. In equ. (6.3) the term $[1+\exp (E/kT)]$ arises from the Fermi distribution function and leads to a sudden increase of $i(E)$ for $E=E_F$. The second exponential factor is due to the energy dependence of the tunneling probability which decreases rapidly for energies below E_F. The result is drawn schematically in fig. 6.2, from which it is evident that only electrons in the range of 0 to about 2 eV below E_F will contribute to the emission current, so that only information about the electronic structure within this limited energy range can be expected by this technique. The slight tailing of $i(E)$ for E above E_F arises from the finite temperature of the solid.

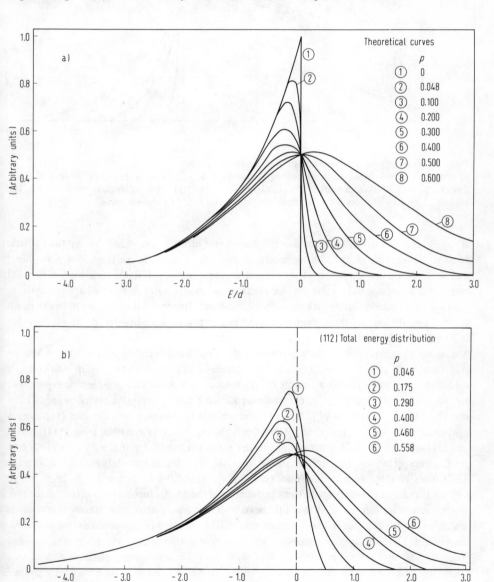

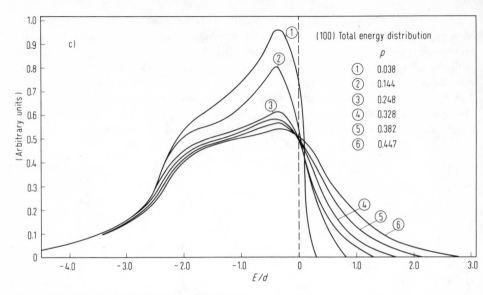

Fig. 6.3. Total-energy-distribution curves from tungsten.
a) Theoretical curves based on the free-electron model [7]. b) TED from W(112). c) TED from W(100).
The energy scale is taken in units of $d \equiv \hbar e F/2(2m\,\phi)^{1/2}\,t(y)$, $t(y)$ being an image correction term. The
parameter $p = kT/d$ denotes the influence of the temperature. After Swanson and Crouser [11].

If the electrons in a solid are not described by a free electron gas model but by the realistic
band structure, equ. (6.3) has to be modified, since the Fermi surface is no longer a sphere,
and depends on the direction of movement of the electrons. Calculations by Stratton [8]
revealed that for normal metals the free-electron gas model is not such a bad approximation,
but that for transition metals with partially filled d-bands stronger deviations are to be expected.
Further calculations on the influence of the band structure have been made by Gadzuk [9]
and by Nagy and Cutler [10].
A number of experimental determinations of the TED have been made. Young and Müller
[12] found fairly good agreement between the measured TED from a tungsten tip with Young's
theoretical prediction [7]. However the energy resolution was not very good in this experiment
and the emission current was integrated over all the emitting crystallographic planes of the
tip. The TED of individual W planes were determined by Swanson and Crouser [11] using a
probehole technique as described by van Oostrom [5]. Whereas emission from (111), (310),
and (211)-planes (fig. 6.3b) corresponds to the predicted behaviour (fig. 6.3a), the (100) and
(110) planes exhibited definite deviations from the free-electron model. Fig. 6.3 shows the
TED from W (100) to have a pronounced 'hump' about 0.4 eV below Fermi level, which
obviously must be connected with some peculiarity of the band structure in the $\langle 100 \rangle$ direction.
Further investigation revealed that this peak is extremely sensitive to surface contaminants
which lead to the conclusion that it represents electron emission from a surface state or surface
resonance [13]. Recently photoelectron spectroscopy was used to measure energy distributions
from W(100) surfaces again revealing this peak, with the same energy position and surface
sensitivity as observed by field emission. This may well be the first experimentally proved
example of the existence of electronic surface states at a clean metal surface.

This conclusion is further supported by Czyzewski [15] who showed that the peak in the TED of W(100) cannot be explained alone by the existence of an enhanced density of states in the bulk.

In conclusion, it can be said that microscopic analysis of TED data from clean surfaces has two main drawbacks, which are that only a very limited range of energies is accessible and that due to the externally applied electric field the emitter surface is not in a state of equilibrium, which may well account for the method's sparse following.

6.3. Adsorbed layers and resonance tunneling

Several interesting features are to be found when field emission from adsorbate covered surfaces is studied.

In a theoretical paper Duke and Alferieff [16] demonstrated that the field emitted current may be enhanced by the presence of adsorbed particles by a process called 'resonance tunneling'. In their model the adsorbed particle is represented by an additional rectangular potential profile in front of the metal surface. The situation is illustrated in fig. 6.4. Neglecting temperature effects the TED of the clean metal (fig. 6.4a) is characterized by an exponential decay, and equ. (6.3) is reduced to

$$i(E) = i_0 \exp(E/\beta) \tag{6.4}$$

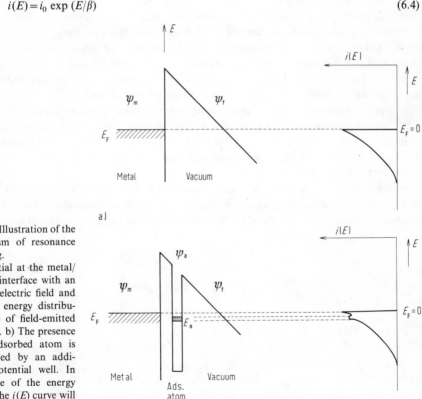

Fig. 6.4. Illustration of the mechanism of resonance tunneling.
a) Potential at the metal/vacuum interface with an external electric field and resulting energy distribution $i(E)$ of field-emitted electrons. b) The presence of an adsorbed atom is represented by an additional potential well. In the range of the energy level E_a the $i(E)$ curve will exhibit a peak.

If in the range of accessible energies an electronic level E_a belonging to an adsorbed particle exists such as that shown in fig. 6.4b, an electron from the metal may tunnel through the first barrier into the adsorbed atom, passing without damping through the atom and finally being emitted with another tunneling probability into the vacuum. Since in case b) the total distance for tunneling between the metallic state at E_a and vacuum is smaller than in case a) the emitted current density will show an increase in this energy region. The magnitude of this effect may be described by an enhancement factor

$$R(E) = i(E)/i_c(E) \qquad (6.5)$$

where $i_c(E)$ is the TED from the clean surface and $i(E)$ that from the adsorbate covered surface. $R(E)$ was evaluated to be between 1 and 10^4. A quantum mechanical calculation is based on a consideration of the corresponding wave functions (fig. 6.4). In the case of a clean surface the tunneling probability, and therefore the emission current, is determined by the magnitude of the overlap intergral $\langle \psi_m | \psi_f \rangle$, whereas for an adsorbate covered surface this term has to be replaced by $\langle \psi_m | \psi_a \rangle \cdot \langle \psi_a | \psi_f \rangle$. Duke and Alferieff [16] gave an exact solution for this problem based on the described one-dimensional model potential.

A somewhat different approach was developed by Gadzuk [17]. He replaced the assumed model by a more realistic description of the potential at a surface created by the adsorption of an atom. His picture correlates directly with the adsorption of alkali metals on transition metals, since in these cases the ionization potential is smaller than the work function ϕ of the substrate. If the atom approaches the surface the ground state level shifts in energy and is broadened due to the interaction with the valence electrons of the substrate metal.

Other model calculations to describe field emission through adsorbed layers were made by Penn et al. [18], Caroli et al. [19], and by Duke and Fauchier [20]. In the latter case again a model potential enabling an exact solution of the problem was used.

Shortly after the theoretical prediction of the effect of resonance tunneling [16] Clark and Young [21] observed experimentally that the field electron emission from an area consisting of about 30 tungsten atoms is increased by a factor of about 3–5 by the presence of a single adsorbed strontium atom, showing that emission through such an adsorbed atom is about one hundred times more effective than from the bare surface.

Plummer et al. [22] measured the enhancement factor $R(E)$ for a W(110) surface covered by zirconium. The results are shown in fig. 6.5 and are in fairly good agreement with the theoretical prediction of Gadzuk [17]. Since $R(E)$ is determined by a combined action of the

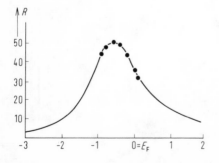

Fig. 6.5. Enhanced tunneling probability R as a function of energy for Zr on W(110). The full line corresponds to the theoretical prediction by Gadzuk [17]. After Plummer et al. [22].

width of the potential well of the adsorbed atom and of the energy width and shift, these factors may not be determined separately, but an estimate of these quantities was derived.

Plummer and Young [24] measured the TED of tungsten surfaces clean and covered with Ba, Sr, or Ca using an energy analyzer with extremely high resolution (some 10 meV). Fig. 6.6a

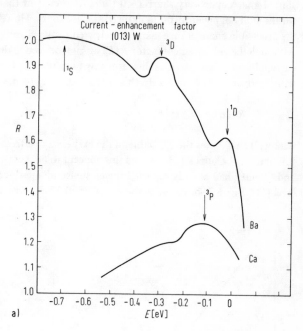

Fig. 6.6. a) Experimental enhancement factor R for Ba and Ca on W(013) as a function of energy ($E_F = 0$).

a)

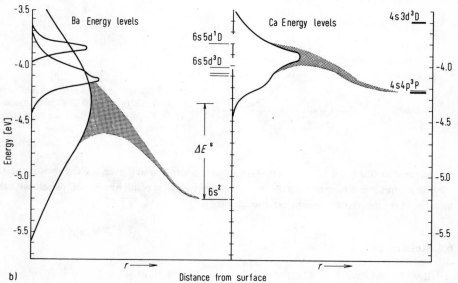

b)

b) Illustration of the shifting and broadening of the energy levels of Ba and Ca on tungsten. After Plummer and Young [24].

shows the results for a W(013) surface with adsorbed Ba and Ca. The pronounced maxima in the $R(E)$ curves are transformed in fig. 6.6b into an energy level diagram representing the adsorbed states. The ground state level of Ba is shifted by 0.95 eV and broadened to a width of 0.75 eV thus leading to some overlap with the excited 3D and 1D states, which are not shifted and remain very sharp (~ 0.1 eV). It is evident that this technique is very promising for the study of chemisorption induced surface states. However as in the case of clean surfaces the accessible energy range again is restricted to the vicinity of the Fermi level.

Although the enhancement factor $R(E)$ was introduced to describe the increase of the emission current by adsorbed particles this term may formally also be used to describe the extra emission from intrinsic surface states. In this case $R^*(E)$ may be defined by

$$R^*(E) = i(E)/i^*(E) \tag{6.6}$$

where $i^*(E)$ denotes the TED due to the existence of a free electron gas as given by equ. (6.3). Plummer and Gadzuk [23] applied this concept to field emission from W(100) where Swanson and Crouser had already reported the existence of a surface state at 0.37 eV below the Fermi level [11]. Fig. 6.7 shows the variation of R^* (0.37 eV) with the exposure of the surface to CO.

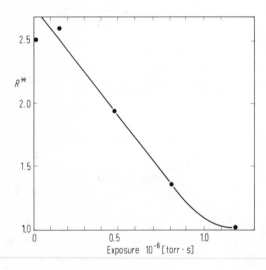

Fig. 6.7. Variation of the enhancement factor R^* for W(100) at $E = -0.37$ eV with exposure to CO. $T = 78$ K. After Plummer and Gadzuk [23].

As can be seen this feature is very sensitive to gas adsorption and even a CO exposure of 1 L leads to a complete disappearance of this extra emission. This result agrees completely with the findings of recent photoemission experiments [14].

6.4. References

[1] E. W. Müller, Phys. Z. 37, 838 (1936); Z. Phys. 106, 541 (1937) (FEM).
[2] E. W. Müller and T. T. Tsong: Field Ion Microscopy. Elsevier, New York 1969.
[3] R. Gomer: Field Emission and Field Ionization. Harvard University Press 1961.
[4] K. M. Bowkett and D. A. Smith: Field Ion Microscopy. North Holland, Amsterdam 1970.

[5] A. van Oostrom, Philips Res. Rept. Suppl. *11*, 102 (1966).

[6] R. H. Good and E. W. Müller in: Handbuch der Physik. (Ed. S. Flügge), Springer-Verlag, Berlin 1956, Vol. XXI, p. 176.

[7] R. D. Young, Phys. Rev. *113*, 110 (1959).

[8] R. Stratton, Phys. Rev. *135*, A 794 (1964).

[9] J. W. Gadzuk, Phys. Rev. *182*, 416 (1969).

[10] D. Nagy and P. H. Cutler, Phys. Rev. *186*, 651 (1969).

[11] L. F. Swanson and L. C. Crouser, Phys. Rev. *163*, 622 (1967).

[12] R. D. Young and E. W. Müller, Phys. Rev. *113*, 115 (1959).

[13] E. W. Plummer and J. W. Gadzuk, Phys. Rev. Lett. *25*, 1493 (1970).

[14] a) J. Waclawski and E. W. Plummer, Phys. Rev. Lett. *29*, 783 (1972).
     b) B. Feuerbacher and B. Fitton, Phys. Rev. Lett. *29*, 786 (1972).

[15] J. J. Czyzewski, Surface Sci. *33*, 589 (1972).

[16] C. B. Duke and M. E. Alferieff, J. Chem. Phys. *46*, 923 (1966).

[17] J. W. Gadzuk, Phys. Rev. *B1*, 2110 (1970).

[18] D. Penn, R. Gomer and M. H. Cohen, Phys. Rev. Lett. *27*, 26 (1971).

[19] C. Caroli, D. Lederer and D. Saint James, Surface Sci. *33*, 228 (1972).

[20] C. B. Duke and J. Fauchier, Surface Sci. *32*, 175 (1972).

[21] H. E. Clark and R. D. Young, Surface Sci. *12*, 385 (1968).

[22] E. W. Plummer, J. W. Gadzuk and R. D. Young, Solid State Comm. *7*, 487 (1969).

[23] E. W. Plummer and J. W. Gadzuk, Phys. Rev. Lett. *25*, 1493 (1970).

[24] E. W. Plummer and R. D. Young, Phys. Rev. *B1*, 2088 (1970).

[25] J. W. Gadzuk and E. W. Plummer, Rev. Mod. Phys. *45*, 487 (1973).

7. Ion Neutralization Spectroscopy

Hagstrum [1, 2] at the Bell laboratories has developed a method during the last few years which is capable of giving information about the energies of electronic states at the surface. This method is superior to photoelectron spectroscopy in so far as the detected electrons originate only from the surface layer whereas with XPS and UPS a portion of the electrons stems from deeper layers. Field emission spectroscopy is similarly surface sensitive but ion neutralization spectroscopy (INS) is capable of covering a larger energy range (about 10 eV). The disadvantage of INS is the complexity involved with the experiments and in calculating electron energy distributions (or some similar functions) from the measured data. Therefore this technique has found only limited use hitherto.

The mechanism of INS consists of a two-electron process which results from the neutralization of a slow noble gas ion at a surface. The energy liberated is transferred to a second electron in an Auger process which can then be ejected from the solid.

The mechanism is illustrated by fig. 7.1. If an ion approaches the surface, an electron (1)

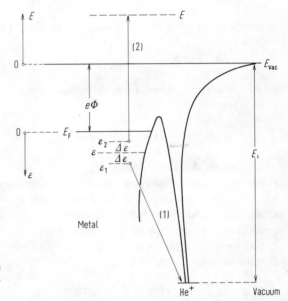

Fig. 7.1. Energy diagram illustrating the mechanism of ion neutralization spectroscopy.

tunnels through the potential barrier and drops into the vacant level of the noble gas atom. A second electron (2) may leave the solid with the corresponding energy. Since the initial energies of electrons (1) and (2) may be anywhere within the filled valence band a continuous distribution of electrons will be excited, a part of which may escape from the solid and whose kinetic energy distribution is measured. The only purpose of the gaseous ion is to offer an unfilled electron state with defined energy. In order to obtain useful information of the electron distribution at the solid surface the approaching ions must not have any states within the energy range of interest in the solid. In most cases slow He^+ ions are used in the experiments.

Measurements with different kinetic energies of the ions allow the results to be extrapolated to give the smallest energy broadening.

From fig. 7.1 it can be seen that the energy gained by electron (1) equals $E_i - \varepsilon_1$ which is transferred to electron (2) so that

$$E + e\phi + \varepsilon_2 = E_i - \varepsilon_1$$

or

$$E = E_i - e\phi - \varepsilon_1 - \varepsilon_2$$

(7.1)

E_i is the energy of the ground state of the noble gas atom at the surface (which differs somewhat from the value for the free atom [2]). ϕ is the work function of the sample.

If at first it is assumed that the transition probability is constant, that means independent of the energy ε, then the probability for the process is proportional to $N(\varepsilon_1) \cdot N(\varepsilon_2)$, $N(\varepsilon)$ being the density of filled electronic states. The probability for excitation of electrons with energy E is then given by

$$\int_{-\varepsilon}^{+\varepsilon} N(\varepsilon - \Delta\varepsilon) \cdot N(\varepsilon + \Delta\varepsilon) \, \mathrm{d}(\Delta\varepsilon)$$

(7.2)

where the energy ε is located halfway between E and E_i on the energy scale, i.e.

$$E = E_i - e\phi - 2\varepsilon$$

(7.3)

Consideration of variations of the transition probabilities between initial and final state leads to an expression of the form

$$\int_{-\varepsilon}^{+\varepsilon} |H_{fi}|^2 N(\varepsilon - \Delta\varepsilon) \cdot N(\varepsilon + \Delta\varepsilon) \, \mathrm{d}(\Delta\varepsilon)$$

(7.4)

which includes the matrix element H_{fi}.

Further complicating factors are the final state density and final state interactions, problems which are similar to those in UPS. These parameters (which are more or less unknown) may be included in a function $N^*(\varepsilon)$ which is called the transition density for the individual electrons involved and which replaces the density of states $N(\varepsilon)$. Then $Y(\varepsilon)$ which is the transition density for pairs of electrons appropriate to processes producing excited electrons at energy E, is the self-convolution of $N^*(E)$:

$$Y(\varepsilon) = \int_{-\varepsilon}^{+\varepsilon} N^*(\varepsilon - \Delta\varepsilon) \cdot N^*(\varepsilon + \Delta\varepsilon) \, \mathrm{d}(\Delta\varepsilon)$$

(7.5)

(Note that this result bears some resemblance to APS, which is also characterized by a two-electron process). $Y(\varepsilon)$ can be simply transformed to $Y(E)$ by means of eq. (7.3). This internal distribution of excited electrons $Y(E)$ is related with the distribution of externally observed Auger electrons $X(E)$ through the probability of electron escape over the surface barrier $P(E)$:

$$X(E) = Y(E) \cdot P(E)$$

(7.6)

$X(E)$ is the measured quantity from which the transition density $N^*(E)$ (which reflects the density of states $N(E)$) has to be evaluated.

The experimental set-up as used by Hagstrum [3] is shown in fig. 7.2. Noble gas ions are produced by electron impact and are focussed onto the target through a lens system. The energy distribution of the ejected electrons is evaluated from the slope of the current drawn by the spherical collector as a function of retarding potential. An improvement was achieved by using a hemispherical grid in front of the collector which prevents variations of the ion focus conditions during the potential sweep of the collector. The instrumental resolution is about 0.1 eV at an electron energy of 10 eV. Noise is suppressed using a multichannel scaler. The apparatus is further equipped with provisions for LEED and photoelectron spectroscopy as well as for cleaning the sample surface by ion sputtering.

After the experimental determination of the electron distribution $X(E)$ this function has to be divided by a parametric $P(E)$ function [2] in order to obtain $Y(E)$. Data in the range $0 < E < 2\,\text{eV}$ are not used since $P(E)$ and $X(E)$ vary rapidly within this range.

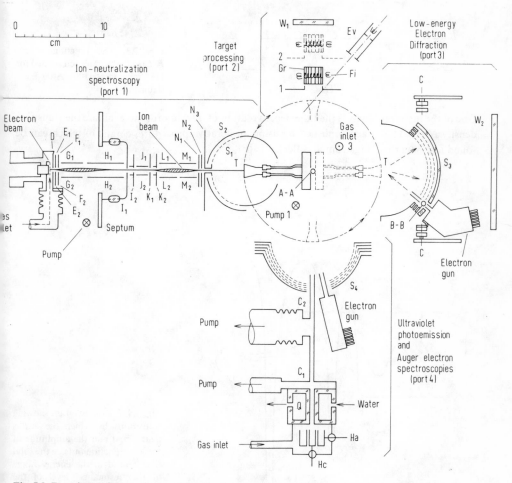

Fig. 7.2. Experimental set-up for ion neutralization spectroscopy. The apparatus includes further facilities for LEED, UPS and AES. The sample T may be rotated around axis A-A. After Hagstrum [2].

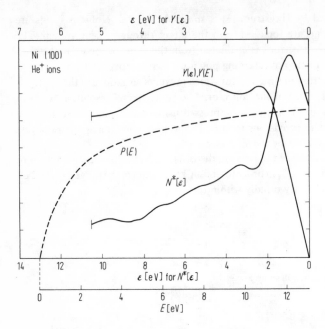

Fig. 7.3. The functions $Y(E)$, $P(E)$ and $N^*(E)$ as determined for He$^+$ ions on Ni(100). After Hagstrum [1].

One of the main problems is then the deconvolution of $Y(\varepsilon)$ in order to evaluate the transition density $N^*(\varepsilon)$. This is accomplished by using an iterative procedure, details of which can be found in a paper by Hagstrum and Becker [4].

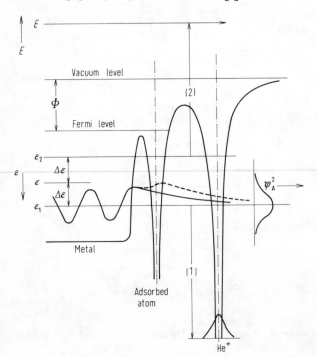

Fig. 7.4. Energy diagram showing transitions by which the ion is neutralized and the amplitude of the wavefunction outside the solid is enhanced in the energy range of the 'resonance' level of the adsorbed atom. After Hagstrum [1].

Results on clean surfaces have so far been obtained for Cu, Ni, Si, and Ge [1]. As an example fig. 7.3 shows the functions $P(E)$, $Y(E)$ and $N^*(E)$ for Ni(100) [5]. In the $N^*(\varepsilon)$ curve the peaked d-band just below the Fermi level E_F ($\varepsilon=0$) is clearly discernible from the broad sp-band. The d-band is somewhat narrower than in other (bulk) studies, indicating some difference of the electronic structure of the surface. Variations between differently oriented single-crystal surfaces are small but well outside the limits of experimental error. This fact is explicable as differences in the electron momenta in directions perpendicular to the surface [5]. There is strong experimental and theoretical [6] evidence that the ejected electrons arise predominantly from the first layer of the solid, so that this method is very sensitive to the existence of chemisorption induced 'resonance' levels at the surface. This situation is illustrated by fig. 7.4. Because the wavefunction of the chemisorption state is expected to have a 'tail' at

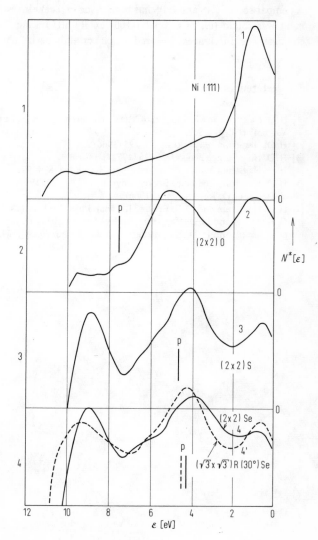

Fig. 7.5. Transition density functions $N^*(\varepsilon)$ for clean Ni(111) and 0, S, and Se adsorbed on Ni(111). After Becker and Hagstrum [8].

the position of the incoming ion this effect leads to an enhancement of the emitted electron current in this energy range.

Results have been obtained for O, S, and Se adsorbed on Ni(100) [7] and on Ni(110) and Ni(111) [8]. These experiments have been combined with LEED observations in order to characterize the state of adsorption. Some of the evaluated $N^*(E)$-curves are reproduced in fig. 7.5. 'p' denotes the energies of p-orbitals in the free chalkogen atoms. It can be seen that the d-band peak at about 1 eV below E_F is decreases by adsorption, whereas new peaks due to the formation of chemisorption-orbitals emerge. There are no significant differences between the three surface orientations.

The authors conclude that oxygen adsorption may well be associated with reconstruction of the surface atoms. In cases where no reconstruction takes place, there is some evidence that bonding occurs to as many Ni atoms as possible [8].

A comparison of INS and UPS has been made in recent investigations [9]. In some cases agreement was found (e.g. CO on Ni(100) exhibiting a state 7.8 eV below E_F) but sometimes characteristic differences appeared which certainly need further elucidation.

7.1. References

[1] H. D. Hagstrum in: Techniques of Metals Research. (Ed. R. F. Bunshah), John Wiley, New York 1972, Vol. 6. Part 1. p. 309.
[2] H. D. Hagstrum, Phys. Rev. *150*, 495 (1966).
[3] H. D. Hagstrum, Science *178*, 275 (1972).
[4] H. D. Hagstrum and G. E. Becker, Phys. Rev. *B4*, 4187 (1971).
[5] H. D. Hagstrum and G. E. Becker, Phys. Rev. *159*, 572 (1967).
[6] V. Heine, Phys. Rev. *151*, 561 (1966).
[7] H. D. Hagstrum and G. E. Becker, J. Chem. Phys. *54*, 1015 (1971).
[8] G. E. Becker and H. D. Hagstrum, Surface Sci. *30*, 505 (1972).
[9] G. E. Becker and H. D. Hagstrum, J. Vac. Sci. Techn. *10*, 31 (1973).

8. Work function and contact potential

8.1. Introduction

Adsorption is usually associated with a variation of the electron work function ϕ by values ranging from a few tenths to more than 1 eV. Since this quantity can be measured easily with a high degree of accuracy (~ 1 meV) the work function provides a simple procedure for monitoring the state of a surface. Moreover the sign of the work function change $\Delta\phi$ is connected with the direction of the electron transfer between the substrate and the adsorbate, i.e. with the dipole moment of the adsorbate complex. In addition $\Delta\phi$ can frequently be used as a measure of the degree of coverage, although this relation needs not necessarily be linear and must be calibrated by independent methods in every case. On the other hand the theoretical treatment of the work function and its variation are still in a rather rudimentary stage so that quantitative information about charge distributions cannot be reliably derived from work function measurements at present.

The work function ϕ of a solid is defined by the energy $e\phi$ which is necessary to transfer an electron from the Fermi energy to the vacuum level, i.e. to a point in infinity outside the crystal. (Fig. 8.1). This quantity is also related to the electrochemical potential $\bar{\mu}$ of the electrons in the solid by [1]

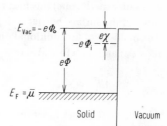

Fig. 8.1. Potential energy diagram for the solid/vacuum interface.

$$e\phi = -e\phi_0 - \bar{\mu} \tag{8.1}$$

where ϕ_0 is the electrostatic potential just outside the surface. Frequently $-e\phi_0 = E_{vac}$ is taken as the origin of the energy scale.

Some modification of the above definition is necessary in the presence of an external accelerating electric field [2].

The electrochemical potential $\bar{\mu}$ is composed of the chemical potential μ of the electrons in the solid (a quantity which is independent of the electrostatic potential) and a contribution from the electrostatic potential ϕ_i inside the solid:

$$\bar{\mu} = \mu - e(\phi_0 + \phi_i) \tag{8.2}$$

so that the work function contains these contributions as well

$$\phi = -\mu/e + (\phi_i - \phi_0) \tag{8.3}$$

The second term is caused by the asymmetric distribution of the electronic charge at a surface since the density of electrons spreads somewhat beyond the limits of the bulk because of the kinetic energy of the electrons [1]. As a consequence an electric dipole layer is built up at the clean surface corresponding to a potential difference $\chi = \phi_0 - \phi_i$ where ϕ_0 ($=0$) is the electrostatic potential of an electron just outside the surface. This electric double layer depends on the geometry of the atomic positions and therefore also on the surface orientation. This is the reason why different crystal planes have different work functions.

In the case of a polycrystalline surface consisting of patches with different crystallographic orientation the electrostatic potentials outside the surfaces are also different, leading to local electric fields. But at a distance which is large compared with the patch dimensions the potential attains a constant value $\bar{\phi}_0$ given by [1]

$$\bar{\phi}_0 = \sum_j \alpha_j \phi_{0,j} \tag{8.4}$$

where α_j is the fraction of the total area covered by the surface of type j. Hence also a mean work function $\bar{\phi}$ arises, given by

$$\bar{\phi} = \sum_j \alpha_j \phi_j \tag{8.5}$$

ϕ_j being the work function of surface j. In the case of large external electric fields (like in field emission experiments) this mean work function loses its meaning and the passage of electrons through the surface barrier is dependent only on the local work function [2].

The build-up of an additional dipole layer at the surface leads to a variation of the difference $(\phi_i - \phi_0)$ and therefore also to a variation of the work function which may be defined as

$$\Delta \phi = \phi_a - \phi_c \tag{8.6}$$

where ϕ_a is the work function of the adsorbate covered surface and ϕ_c that of the clean surface. The surface potential (s.p.) ΔV is a frequently used alternative convention, where

$$\Delta V = \phi_c - \phi_a = -\Delta \phi \tag{8.7}$$

If the surface is not uniformly covered by the adsorbate the problem must be treated as for polycrystalline surfaces.

When two conductors A and B with electrochemical potentials $\bar{\mu}_A$ and $\bar{\mu}_B$ are brought into electrical contact, $\bar{\mu}_A$ and $\bar{\mu}_B$ will change until at equilibrium $\bar{\mu}_A = \bar{\mu}_B$, i.e. the Fermi energies of both materials are at the same level. This is achieved by electron transfer until $(\phi_A - \phi_B)_i$ balances the difference $\mu_A - \mu_B$, the chemical potentials μ being insensitive to the electrostatic potential. The situation is illustrated by fig. 8.2. Therefore it follows that

$$\phi_A + \phi_{0,A} = \phi_B + \phi_{0,B}$$
$$V_{AB} = (\phi_{0,A} - \phi_{0,B}) = \phi_B - \phi_A \tag{8.8}$$

Hence an electrical potential difference V_{AB} called the Volta potential or contact potential difference (CPD) appears between points just outside the surfaces of the two conductors which is equal to the difference in the work functions $\phi_B - \phi_A$. If A is some reference material

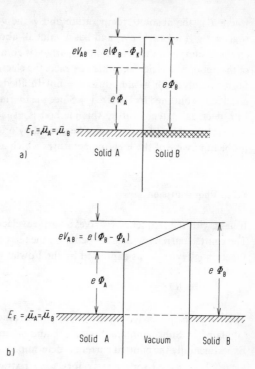

Fig. 8.2. Potential energy diagram for two solids in electrical contact illustrating the build-up of the contact potential difference (CPD) V_{AB}. a) the region at the junction, b) the situation at the surfaces.

whose work function is known not to change during an experiment variations of the work function of B are measured as changes in the contact potential difference:

$$\delta\phi_B = \delta V_{AB}.$$

8.2. Experimental techniques

A variety of methods exists for measuring the work function and its changes which may be classified into those which allow an absolute determination of ϕ and others by which only differences $\Delta\phi$ with respect to a reference state may be measured. Only a few of the most important techniques are outlined here, and for a more detailed discussion the reader is referred to one of the review articles on this topic, e. g. that by Rivière [3].

8.2.1. Thermionic electron emission

The current density i of thermally emitted electrons is described by the Richardson-Dushman equation as

$$i = A T^2 \exp\left(-e\phi/kT\right) \tag{8.9}$$

where T is the absolute temperature and ϕ the work function of the solid. The constant A is given by $A = 4\pi m k^2 e/h^3$ and has a value of about 120 [A/cm^2·K^2]. This equation is based on the assumption that the emitting surface is completely homogeneous and that the influence of the applied electric field which extracts the electrons from the hot cathode can be neglected. Since usually both assumptions are not fulfilled for practical purposes eq. (8.9) has to be modified as outlined by Rivière [3]. Since the thermally emitted current decreases exponentially with decreasing temperature this method is usually only suitable for the refractory metals. The state of the surface must not be altered by or during the measurements which is a serious problem in view of the high temperatures which are needed.

8.2.2. Photoemission

If light with the frequency v strikes a surface, electrons will be emitted provided that $hv > e\phi$. The emitted current density i depends on the temperature T, the light frequency v and the work function ϕ ($= hv_0/e$) as expressed by the Fowler equation

$$i = BT^2 \cdot f[(hv - hv_0)/kT] \tag{8.10}$$

The factor B is nearly constant near the threshold frequency v_0. The function f has been tabulated by Suhrmann and Simon [4], and others.

Experimentally the emitted current is determined as a function of v which is plotted in the form $\log (i/T^2)$ vs. hv/kT (or i vs. $\sqrt{v}$ for $v \gg v_0$). Extrapolation to zero current then yields the threshold frequency v_0 and thereby the work function ϕ. Details may be found in the literature [3], [4].

A common feature of both techniques described so far is that in the case of polycrystalline surfaces they do not yield the average work function $\bar{\phi}$ as defined by equ. (8.5) but that the derived values are dominated by the patches with the lowest local work function even if these cover only a very small fraction of the total surface area.

8.2.3. Field emission

An electric field F [V/cm] applied externally to a solid enhances the probability of electron escape via tunnel effect. The emitted current density is given by the Fowler-Nordheim equation [5]

$$i = \frac{e^3 F^2}{8\pi h \phi t^2(y)} \cdot \exp\left[-\frac{4(2m)^{1/2} \phi^{3/2}}{3 h e F} \cdot f^*(y) \right] \tag{8.11}$$

The functions $t(y)$ and $f^*(y)$ (where $y \propto F^{1/2} \phi^{-1}$) are correction terms which deviate only slightly from unity. Therefore a plot of $\log (i/F^2)$ vs. $1/F$ should give a straight line and the work function can be determined by the slope of this line.

The electric field strength must be of the order of 10^7 to 10^8 [V/cm] in order to achieve measurable emission currents, so that this technique is restricted to surfaces with very small radii of curvature ($F \propto U/r$) of about 1000 Å, such as very thin wires or the sharp tips used in field emission microscopy. The accuracy obtainable is rather disappointing and again the patches with the lowest work function dominate. On the other hand the advantages of the field emission microscope may be used to study work functions of individual crystal planes of the same tip by the use of a probe-hole technique [6].

8.2.4. Field emission retarding potential method (FERP)

The preceding techniques are based on measurements of the work function of an emitter and exhibit a series of limitations which are partly also of theoretical nature. Many of the difficulties may be circumvented by a technique which measures the work function of an electron collector. This may be achieved by the FERP method as introduced originally by Henderson and Badgley [11], but which so far has found only a few applications [20]. The theoretical and experimental details have been discussed recently by Strayer et al. [12]. The principle is outlined by fig. 8.3: Electrons emitted from a field emission tip will have (at low temperatures) energies

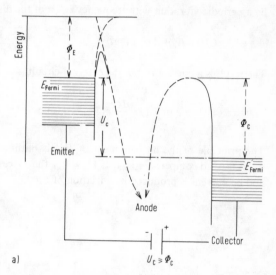

Fig. 8.3. a) Potential energy diagram for the FERP method.

a)

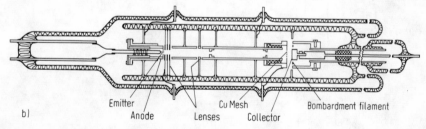

b) Diagram of a FERP tube. After Strayer et al. [12].

very close to the Fermi energy E_F of the emitter, i.e. $-eV_c$ with respect to the Fermi level of the collector. If $V_c = \phi_c$ a sudden increase of the collected current should take place, thus the determination of the threshold voltage yields the work function of the collector.

8.2.5. Vibrating capacitor method (Kelvin method)

The following two techniques are based on the measurement of contact potential differences, i.e. no absolute determination of work functions is possible but only of their changes during variations of the surface state. On the other hand these techniques are not restricted to special shapes or temperatures of the sample and yield results with a high degree of accuracy (1 mV). Therefore they are invaluable to all kinds of adsorption experiments.

If the two conductors are arranged as a capacitor with a small distance between the plates (<1 mm) as illustrated by fig. 8.1 then the existing potential difference $V_{AB} = \phi_B - \phi_A$ just outside the two surfaces leads to a charge on the capacitor

$$Q = C \cdot V_{AB} \tag{8.12}$$

The capacity C is given by the geometry of the arrangement and may be varied for example by a periodic vibration with frequency $\omega/2\pi$ of the distance between the plates:

$$C(t) = C_0 + \Delta C \sin \omega t \tag{8.13}$$

The result is a current $i(t)$ in the external circuit given by

$$i(t) = \frac{dQ}{dt} = V_{AB} \cdot \frac{dC}{dt} = V_{AB} \omega \Delta C \cos \omega t \tag{8.14}$$

The principle of the measurement of V_{AB} is outlined in fig. 8.4. An external voltage source V_{ex} is used to compensate V_{AB}. If $V_{ex} = V_{AB}$ the space between the capacitor plates is field-free and no current is produced by variation of C.

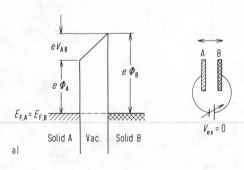

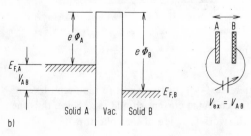

Fig. 8.4. Principle of CPD measurement by the vibrating capacitor (Kelvin) method.
a) Potential energy diagram for $V_{ex} = 0$. b) If $V_{ex} = V_{AB}$ the space between the two plates is field-free.

A self-compensating device [7] is shown schematically in fig. 8.5: The fast electrometer amplifier with an input resistance of about 10^{10} Ω serves as impedance transformer. The ac signal with frequency $\omega/2\pi$ passes through a filter and is received by a lock-in amplifier which is tuned to the frequency $\omega/2\pi$. If the switch is in position 2 then the output of the lock-in amplifier delivers a signal proportional to V_{AB}; by changing the voltage V_{ex} the potential difference between A and B may be varied. With the switch in position 1 the system is self-compensating. The output voltage of the lock-in amplifier is fed back to the condenser until no signal is produced, i.e. the field between the plates is zero. If the work function of the sample changes (e.g. after adsorption) the compensating voltage needed to restore field-free conditions equals $\Delta\phi$ and may be directly plotted as a function of time with a recorder.

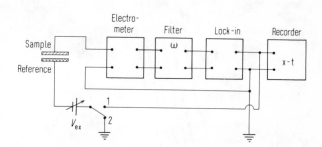

Fig. 8.5. Schematic circuit for a self-compensating Kelvin method [7].

It is important to notice that the work function of the reference electrode must not be varied by the interaction with gases under investigation etc. Gold, oxidized tungsten and oxidized tantalum have been found to be suitable materials.

Experimental details of the capacitor method have been reviewed by Craig and Radeka [8].

8.2.6. Diode method

If the device of fig. 8.2 is considered as a diode where electrons emitted from the hot cathode are collected by the sample surface A acting as anode, then it can be shown that the anode current i_a of this diode operated either in the retarding field or space charge limited mode is only a function of the difference between the applied voltage V_a and of the work function of the anode (!) ϕ_A [9]. A typical series of plots i_a vs. V_a is shown in fig. 8.6. If the work function of the anode is varied by adsorption then the current/voltage curve is shifted parallel to the V-axis by the amount of $\Delta\phi$. It is also possible to operate this technique automatically, for example by using a constant current source which regulates the voltage V_c in order to keep i_a constant [9]. If the sample surface is small it is more convenient to use an electron gun instead of an isotropically emitting filament. This may be achieved by the electron source of a LEED system as described by Chang [10] and widely used in LEED work.

An excellent review of the different aspects of the diode method has recently been given by Knapp [2].

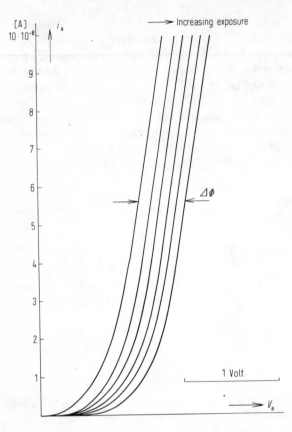

Fig. 8.6. Current/voltage curves for the diode method applied to CO adsorption on Pd(100). The electron gun of a LEED system served as cathode.

8.3. The work function of clean surfaces

Several attempts have already been made at a theoretical description of the work function ϕ. The simplest approach [13] correlates ϕ with the electronegativity X_a of the atoms by an empirical equation:

$$\phi \approx 0.817 \cdot X_a + 0.34 \tag{8.15}$$

This relation consists of two terms, the first one representing the energy required to transfer a valence electron from the atom to infinity and the second describes the influence of the electric dipole layer at the surface. Fig. 8.7 shows the relation between the work functions and electronegativities of a series of metals.

It is easy to understand why not the total amount of X_a enters equ. (8.15), since in the lattice the electrons are delocalized to some extent. Equ. (8.15) gives a fairly satisfactory overall picture but is not able to explain details like the variation of ϕ with the surface orientation for example. Steiner and Gyftopolous [15] have suggested that the electronegativity should in general depend on the crystallographic surface orientation just as the dipole layer is influenced by the atomic density of the plane [16].

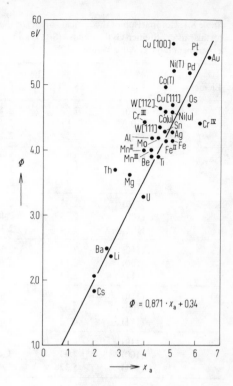

Fig. 8.7. Correlation between work functions ϕ and electronegativities X_a for a series of metals and metal ions (in compounds). After Moesta [14].

Recently Kohn and Lang [17] published a more quantitative description. The calculations were based on explicit evaluations of the charge density 'below' and 'above' the surface yielding the dipole contribution for a simple model (jellium). The bulk contribution (i.e. the chemical potential μ) was derived with the aid of the nearly-free electron model. A detailed description of the theory is given by Lang [18].

Some experimental values are compared with those derived from the Lang-Kohn theory [17] and the semi-empirical data from Steiner and Gyftopoulos [15] in table 8.1. Rivière [3] has given a survey of the work functions of many metals and compounds.

8.4. Adsorbed layers

In nearly all cases of adsorption the work function of the substrate either increases or decreases, which is due to a modification of the dipole layer at the surface. The formation of a chemisorption bond is associated with a partial electron transfer between substrate and adsorbed particles; in the case of physisorption the polarization of the adsorbed particles by the electric field at the surface gives rise to the build-up of a dipole layer. The variation of the work function $\Delta\phi$ is related with the density n_s of adsorbed particles and the dipole moments μ_s of the individual adsorbate complexes. The simplest picture assumes a continuous and infinite double layer of charge separated by a distance d. The total dipole moment per cm^2 is then given by

$$n_s \cdot \mu_s = \varrho \cdot d \tag{8.16}$$

Table 8.1. Comparison of theoretical values for the work functions of some metals after Steiner-Gyftopoulos (S-G) and Lang-Kohn (L-K) with experimental data. After Wagner [19].

Element	Plane	Theory		Experiment (polycrist.)
		S.-G.	L.-K.	
Li	(110)	2.78	2.40	
	(100)	2.61	2.40	2.32
	(111)	2.58	2.30	
Na	(110)	2.52	3.10	
	(100)	2.40	2.75	2.75
	(111)	2.39	2.65	
K	(110)	2.35	2.75	
	(100)	2.25	2.40	2.39
	(111)	2.24	2.35	
Cs	(110)	2.23	2.60	
	(100)	2.14	2.30	2.14
	(111)	2.13	2.20	
Al	(110)	3.65	3.65	
	(100)	3.92	4.20	4.19
	(111)	4.12	4.05	
Cu	(110)	4.65	3.55	4.48
	(100)	4.99	3.80	4.59
	(111)	5.32	3.90	4.94

where ϱ is the charge density per square centimeter. On the other hand the capacity per unit area $C = \varepsilon_0/d$ is related to the potential drop $\Delta\phi$ through

$$\Delta\phi \cdot C = \varrho \tag{8.17}$$

from which follows

$$\Delta\phi = n_s \mu_s / \varepsilon_0 \quad \text{or} \quad \mu_s = \varepsilon_0 \Delta\phi / n_s {}^*) \tag{8.18}$$

Hence $\Delta\phi$ should vary linearly with the coverage θ, provided that the dipole moment μ_s is independent of θ. This condition is, at least at low surface concentrations, frequently fulfilled. However deviations from such a simple proportionality are observed in many cases, some of the reasons of which will be discussed below.

The second assumption involved in eq. (8.18) is that the surface is a continuous layer rather than being composed of discrete dipoles. This effect has been discussed by Gomer [22], who showed that the measured $\Delta\phi$ depends on the method of measurement. Field emission involves electrons tunneling through a potential barrier which is very close to the surface and therefore is sensitive to local variations of the charge distribution. On the other hand the diode or Kelvin methods depend on the value of the electrostatic potential at much greater distances

---

*) Note that the unit of the dipole moment so derived will be [Cb · cm]; 1 Cb · cm = $3 \cdot 10^{27}$ Debye ($\varepsilon_0 = 8.85 \cdot 10^{-14}$ Cb/V · cm).

from the surface. In such cases eq. (8.18) will be valid except possibly at extremely low coverages, i.e. if the mean distance between the adsorbed particles reaches values of the same magnitude as the separation from the reference electrode.

The effect of the mutual influence of the adsorbed dipoles is generally complex and depends on the nature of the adsorption bond. The simplest model which is applicable to physisorbed layers assumes only direct interactions between the dipoles leading to mutual depolarization and thus to a reduction of the individual dipole moments.

This effect was first treated by Topping [23] who derived the following relation between $\Delta \phi$ and the surface concentration n_s:

$$\Delta \phi = \frac{n_s \, {}^\circ \mu_s}{\varepsilon_0 (1 + 9 \alpha n_s^{3/2})} \tag{8.19}$$

$^\circ \mu_s$ is the dipole moment of the adsorbate complex at zero coverage and α is the polarisability of the adsorbed particles. Rearranging eq. (8.19) shows that a plot of $n_s/\Delta \phi$ versus $n_s^{3/2}$ should yield a straight line. This relation was found to hold reasonably well for the Xe/Pd(100) system [24], for example, but the magnitude of α derived from the slope of this line is too large if compared with the value for free Xe atoms. Certainly the Topping formula must at least be modified to take the geometrical arrangement of the adsorbed particles into consideration, such as was done by Carroll and May [25] who found considerable deviations from the simple picture.

If metals with small ionization energies, e.g. the alkali metals, are adsorbed the work function of the substrate metal decreases and often exhibits a minimum in a plot of $\Delta \phi$ against coverage, as shown for example in fig. 8.8 for the system Na/Ni(100) [26]. This effect was first observed

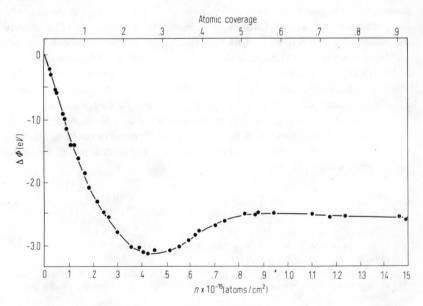

Fig. 8.8. Work function change $\Delta \phi$ as a function of Na coverage on Ni(100). After Gerlach and Rhodin [26].

by Langmuir [27] who tried to explain the phenomenon in terms of a valence electron being transferred from the adsorbed particle to the substrate which creates a dipole consisting of the positive ions and their image charge in the substrate. These dipoles should mutually depolarise giving a minimum in the work function at coverages where the increase of the number of dipoles (=coverage) is just compensated by the decrease of the individual dipole moments due to the depolarisation effect. The Topping formula clearly fails to describe this effect but a qualitatively good description is possible using the semi-empirical relation derived by Gyftopoulos and Levine [23]:

$$\frac{\Delta\phi}{\phi_{\text{sub}}-\phi_{\text{ads}}}=1-G(\theta)\cdot\frac{A\theta}{1+B\theta^{3/2}} \tag{8.20}$$

where ϕ_{sub} is the work function of the bare substrate, ϕ_{ads} that of the bulk phase of the adsorbate. A and B are constants, and $G(\theta)$ an analytical function for which the form $G(\theta)=1-3\theta^2+2\theta^3$ has been found to be favourable.

Gadzuk [29] correlated a quantum mechanical treatment of the metal/metal bond with the variation of the work function. His model is based on the assumption that the level of the valence electrons of the adsorbed atom is shifted in energy and broadened upon approaching the surface of the substrate. The dipole length d depends, of course, on the geometrical arrangement on the surface. From the energy shift and the broadening a density of occupation of the adsorbate level results which determines the net charge. The dipole moments of adsorbed alkaline metals on metal surfaces at low coverages can be evaluated reasonably well by this technique, but a treatment of the mutual interactions which are relevant at higher coverages is not included.

A description of the $\Delta\phi(\theta)$ correlation for such systems was conceived by Lang [30] on the basis of his treatment of the work functions of clean metals [18]. The theoretical results agree fairly well with the overall features of experimental observations. The 'density functional formalism' used differs from Gadzuk's approach in so far as no microscopic quantities (which are almost unknown) are included in the calculations. Thus this treatment has some resemblance to Langmuir's explanation.

Systems with chemisorbed gases are in general even more complex than those with metallic adsorbates because different adsorbed states may occur. As well as simple linear relationships between $\Delta\phi$ and θ, minima have been observed as with metallic adsorbates (e.g. with the system CO/Cu [31]). A theoretical description must of course be based on a quantum chemical treatment of the chemisorption bond [32] and of the mutual and indirect interactions between the adsorbed species [21, 33].

8.5. References

[1] C. Herring and M.H. Nichols, Rev. Mod. Phys. *21*, 185 (1949).
[2] A.G. Knapp, Surface Sci. *34*, 289 (1973).
[3] J.C. Rivière, in: Solid State Surface Science. (M. Green ed.) Vol. I, M. Dekker, New York 1969.
[4] H. Simon and R. Suhrmann: Der lichtelektrische Effekt und seine Anwendungen. Springer-Verlag Berlin 1958.
[5] See for example: R. Gomer: Field Emission and Field Ionization. Harvard University Press, Cambridge, Mass. 1961.

[6] See for example: B.E. Nieuwenhuys and W.M.H. Sachtler, Surface Sci. *34*, 317 (1973); W.M.H. Sachtler, Angew. Chem. *80*, 673 (1968).

[7] G. Ertl and D. Küppers, Ber. Bunsenges. *75*, 1017 (1971).

[8] P.P. Craig and V. Radeka, Rev. Sci. Instr. *41*, 258 (1970).

[9] D.F. Klemperer and J.C. Snaith, J. Phys. E. (Sci. Instr.) *4*, 860 (1971).

[10] C.C. Chang, Thesis, Cornell University 1967.

[11] J.E. Henderson and R.E. Badgley, Phys. Rev. *38*, 590 (1931).

[12] R.W. Strayer, W. Mackie and L.W. Swanson, Surface Sci. *34*, 225 (1973).

[13] W. Gardy and W.J.O. Thomas, J. Chem. Phys. *24*, 439 (1956).

[14] H. Moesta: Chemisorption und Ionisation in Metall-Metall-Systemen. Springer Verlag, Berlin 1968.

[15] D. Steiner and E.P. Gyftopoulos, Proc. 27th Phys. Electronics Conf., Cambridge, Mass., 1967, p. 169.

[16] R. Smoluchowski, Phys. Rev. *60*, 661 (1941).

[17] N.D. Lang and W. Kohn, Phys. Rev. *B3*, 1215 (1971).

[18] N.D. Lang: 'The density-functional formalism and the electronic structure of metal surfaces' in Solid State Physics Vol. 28 (H. Ehrenreich, F. Seitz and D. Turnbull, eds.), Academic Press, New York 1973, p. 225.

[19] H. Wagner, Proc. 3rd Int. Conf. on Thermionic Electrical Power Generation, Jülich 1972.

[20] A.A. Holscher, Surface Sci. *4*, 89 (1966).

[21] T.B. Grimley, Proc. Phys. Soc. (London), *90*, 751 (1967); *92*, 776 (1967).

[22] R. Gomer, J. Chem. Phys. *21*, 1869 (1953).

[23] J. Topping, Proc. Roy. Soc. (London), *A114*, 67 (1927).

[24] P.W. Palmberg, Surface Sci. *25*, 589 (1971).

[25] C.E. Carroll and J.W. May, Surface Sci. *29*, 60 (1972).

[26] R.L. Gerlach and T.N. Rhodin, Surface Sci. *19*, 403 (1970).

[27] I. Langmuir, J. Am. Chem. Soc. *54*, 2798 (1932).

[28] E.R. Gyftopoulos and J.D. Levine, J. Appl. Phys. *33*, 67 (1962).

[29] J.W. Gadzuk, Surface Sci. *11*, 465 (1968).

[30] N.D. Lang, Solid State Comm. *9*, 1015 (1971); Phys. Rev. *B4*, 4234 (1971).

[31] J. Pritchard, J. Vac. Sci. Techn. *9*, 895 (1972).

[32] J.R. Schrieffer, J. Vac. Sci. Techn. *9*, 561 (1972).

[33] T.L. Einstein and J.R. Schrieffer, J. Vac. Sci. Techn. *9*, 956 (1972).

9. Low energy electron diffraction (LEED)

9.1. Historical development

The discovery of interference phenomena in electrons scattered by crystals is closely related to the development of quantum mechanics. In his thesis L. de Broglie [1] postulated the wave nature of matter in 1924, whereafter a flux of particles with velocity v and mass m is correlated with a wavelength $\lambda = \dfrac{h}{mv}$, h being the famous Planck's constant. It follows immediately from this equation that the wavelength for 100 eV-electrons is about 1 Å, so that interference of these waves with periodic crystal lattices is to be expected, just as x-rays with similar wavelength are diffracted.

In 1921 Davisson and Kunsman [2] reported anisotropies in the angular distributions of electrons backscattered from polycrystalline Ni targets. A short time later Farnsworth [3] began experiments on the scattering of slow electrons at metals and interpreted the observed anomalies (which were very probably caused by interference effects) as experimental artefacts. After publication of de Broglie's ideas Elsasser [5] tried to discuss experimental data on Pt from Davisson and Kunsman [4] in terms of diffraction of electron waves.

However, the essential insight was achieved by an accident. In April 1925 Davisson and Germer were experimenting with a polycrystalline Ni sample which was annealed at elevated temperatures from time to time in order to obtain reproducible results. During one such heating cycle the glass apparatus was damaged when a vessel filled with liquid air imploded. The sudden inrush of air caused severe oxidation of the hot Ni crystal, and in order to restore the initial conditions the sample was subsequently reduced by prolonged heating in hydrogen. This procedure obviously caused the growth of larger monocrystalline grains by recrystallization which afterwards gave rise to the appearance of pronounced maxima in the angular distribution of backscattered electrons. The first assumption was that the observed maxima and minima were the result of directional 'crystal transparency' and could be interpreted on the basis of a simple particle model, but more detailed investigations showed unequivocally that this effect was caused by the wave character of electrons. In their paper published in April 1927 entitled "Diffraction of electrons by a crystal of nickel", Davisson and Germer [6] analysed the directions of diffraction maxima on the basis of de Broglie's equation. Even in this first paper they emphasize the need for a dynamic theory, and also discuss the formation of ordered adsorbed structures ("two-dimensional gas crystal"). The importance of temperature effects was also noticed.

Compared with modern systems their apparatus was quite simple and consisted of a glass bulb which was evacuated during bake-out to $2 \cdot 10^{-6}$ Torr and then sealed-off from the vacuum system. A simple charcoal sorption pump cooled with liquid air and the evaporation of some getter metal caused the pressure to fall below the detection limit at this time of $\sim 10^{-8}$ Torr; it can be safely assumed that Davisson and Germer were already working under good ultra high vacuum conditions. The angular distribution of backscattered electrons was registered by a movable Faraday cage which was equipped with a device for eliminating the inelastically scattered electrons by means of a retarding potential.

During the following period the use of diffraction of low energy electrons as a tool for surface investigation was virtually restricted to only a few research groups, in particular Farnsworth and his co-workers. The reason is to be found in the stringent requirements for vacuum conditions and surface structure, as well as in the tedious procedure of detection by means of the movable Faraday cage.

In 1934 Ehrenberg [7] suggested accelerating the diffracted electrons onto a fluorescent screen after passing through a system of fine grids. However this idea which allows the total diffraction pattern to be made "visible", was not developed further until 1960 when Germer et al. [8] initiated a renaissance of the LEED method with their "postacceleration system". Lander et al. [9] constructed a system with spherical grids and fluorescent screen in 1962, and this device is the basis of most of the systems of the "display-type" presently in use. Following an earlier proposal of Sproull [11], Tucker [10] developed a system with magnetic deflection of the incident and diffracted electron beams. In 1964 Park and Farnsworth [12] constructed a completely automated version of the Faraday cup technique which allows exact intensity measurements and a rapid registration of the diffraction pattern on a storage oscilloscope.

The need for a thorough dynamic theory for LEED intensities was emphasized very soon [13] after the publication of the experimental results of Davisson and Germer. Earlier fundamental theoretical papers were published by Bethe [14] and Morse [15]. There is as yet no complete LEED theory although the work of many theoretical physicists has resulted in considerable progress in recent years.

In the meantime hundreds of papers have been published on LEED and it cannot be the aim of this presentation to give a complete review. A bibliography up to 1969 was compiled by Stephens [16] and a series of review and introductory articles as well as some recent extensive contributions on the more theoretical aspects are included in the reference list [17].

9.2. Classification of surface structures

As with three-dimensional crystal structures, periodic two-dimensional surface structures may be classified into certain lattice types. (Deviations from perfect periodicity leading to structural defects are not considered in this context). The surface region (sometimes termed 'selvedge') has indeed a finite thickness but no periodicity in the direction normal to the surface. The 14 Bravais lattices of three-dimensional structures are replaced for surface structures by five types of two-dimensional nets which are formed by equivalent lattice points (fig. 9.1). Identical lattice positions are then connected by translation vectors

$$T = h' \cdot a_1 + k' \cdot a_2 \qquad (h', k' = \text{integers}) \tag{9.1}$$

The periodicity of the surface is usually related to the substrate lattice, i.e. to the periodicity described by unit vectors a_1 and a_2 which is observed when a plane parallel to the surface through the bulk is considered. In other words, the substrate lattice is defined as that plane parallel to the surface below which the three-dimensional (bulk) periodicity is found, and the surface structure as that region above the substrate surface in which lateral periodicity is different and/or in which no periodicity normal to the surface is found [17g]. According to the usual conventions of x-ray crystallography the indexing is chosen so that $|a_1| \leqslant |a_2|$. Connecting the surface periodicity with the substrate (bulk) structure is advantageous because the diffraction spots originating from the substrate periodicity also appear in the LEED pattern and can be readily identified and linked with the known bulk structure.

For most of the clean metal surfaces the surface structure is identical with the substrate

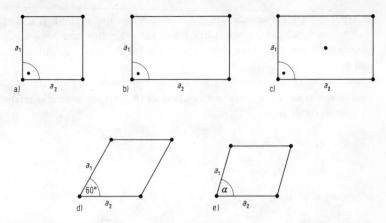

Fig. 9.1. The five types of two-dimensional Bravais lattices
a) square $a_1 = a_2$, $\alpha = 90°$
b) primitive rectangular, $a_1 \neq a_2$, $\alpha = 90°$
c) centered rectangular, $a_1 \neq a_2$, $\alpha = 90°$
d) hexagonal, $\qquad\qquad a_1 = a_2$, $\alpha = 60°$
e) oblique, $\qquad\qquad\ a_1 \neq a_2$, $\alpha \neq 90°$

structure (except for possible vertical displacements of the surface planes), whereas with clean semiconductor surfaces considerable deviations may occur. In particular, variations of the surface periodicity are observed after the formation of ordered adsorbed structures.

The classification of surface structures frequently follows a nomenclature proposed by Wood [18] based on x-ray crystallography. The relation between the surface lattice (with basis vectors b_1 and b_2) and substrate lattice (basis vectors a_1, a_2) is expressed by the ratios of the lengths of the basis vectors $|b_1|/|a_1|$ and $|b_2|/|a_2|$, and any angle of rotation ϑ between the two lattices. If $a_1//b_1$ and $a_2//b_2$ the angle $\vartheta = 0$ is omitted in the notation. The unit cell of a surface structure may either be primitive (p) or centred (c). Following this nomenclature fig. 9.2a shows a p2 × 2 – (or simply a 2 × 2 –) structure and fig. 9.2b a c2 × 2-structure. The structure

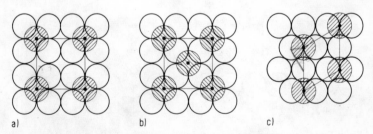

Fig. 9.2. Examples for overlayer structures
a) 2 × 2   b) c(2 × 2)   c) $\sqrt{3} \times \sqrt{3}$ /R 30°.

in figure 9.2c is denoted as $\sqrt{3} \times \sqrt{3}$ /R 30° since both lattices are rotated by 30° with respect to each other.

The advantage of Wood's nomenclature is its simplicity, but unfortunately it cannot describe all types of periodic surface structures adequately. For example it can be seen immediately that it fails in all cases where surface and substrate structure have no common periodicity ('incoherent' structures), or if the angle included by a_1 and a_2 differs from that between b_1 and b_2.

A universally applicable description has been proposed by Park and Madden [19]. The unit cells of substrate (a_1, a_2) and surface lattices (b_1, b_2) are generally related to each other by the transformations

$$b_1 = m_{11} a_1 + m_{12} a_2$$

$$b_2 = m_{21} a_1 + m_{22} a_2 \tag{9.2}$$

or in matrix-notation

$$b = \mathfrak{M} \cdot a \tag{9.3}$$

The structures of fig. 9.2 can for example be described by the matrix $\mathfrak{M} = \begin{pmatrix} m_{11} & m_{12} \\ m_{21} & m_{22} \end{pmatrix}$, where $\mathfrak{M} = \begin{pmatrix} 2 & 0 \\ 0 & 2 \end{pmatrix}$, $\begin{pmatrix} 1 & 1 \\ -1 & 1 \end{pmatrix}$ and $\begin{pmatrix} 1 & 1 \\ -1 & 2 \end{pmatrix}$ for figs. 9.2a)–c), respectively.

The areas for the unit cells of substrate and surface lattices are given by $A = |a_1 \times a_2|$ and $B = |b_1 \times b_2|$. It can be shown that

$$B = (m_{11} m_{22} - m_{22} m_{21}) \cdot A = A \cdot \det \mathfrak{M}. \tag{9.4}$$

All periodic surface structures can now be divided into the following three types on the basis of their relation to the corresponding substrate lattice:

a) If all m_{ij} are integers a and b are simply related to each other. These cases are called simple structures. Also det $\mathfrak{M}$ is an integer. The examples drawn in fig. 9.2 are simple structures.

b) If (a_1, a_2) and (b_1, b_2) are related by rational numbers, then det $\mathfrak{M}$ is a simple fraction and these structures are denoted as coincidence lattices [20]. A one-dimensional example is shown in fig. 9.3a, where $3b_1 = 4a_1$ or $b_1 = \frac{4}{3} a_1$. It is evident that in these cases also a common periodicity of both lattices exists.

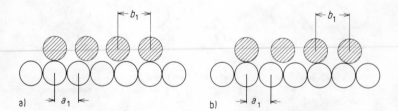

Fig. 9.3. a) One-dimensional example for a coincidince lattice where $3b_1 = 4a_1$.
b) Example for an incoherent structure without a common periodicity between both lattices.

c) An incoherent structure is characterized by irrational numbers relating a and b, which means that also det $\mathfrak{M}$ is an irrational number. No common periodicity between a and b exists in these cases, in contrast to the coincidence lattices. An example is shown in fig. 9.3b.

Incoherent structures are frequently observed for example during the formation of epitaxial layers, where a metal lattice grows on a crystal substrate. Most of the observed ordered adsorbed layers are simple structures corresponding to the earlier concept of the existence of distinct adsorption sites on a surface. Coincidence lattices are a compromise between a configuration given solely by the structure of the substrate (leading to a simple structure) and the most favourable arrangement given by lateral interactions between the surface atoms which would give rise to the formation of an incoherent structure.

As will be shown in detail later there may be a fundamental difficulty in deciding whether an observed diffraction pattern originates from a simple structure ("true superstructure") or from a coincidence lattice with the same periodicity (composed from substrate and surface lattice).

A somewhat different classification of surface structures has recently been proposed by Fingerland [21] on the basis of a common sublattice for the substrate and surface structures *a* and *b*.

If the surface consists of rows of equally spaced steps a nomenclature as proposed by Lang et al. [22] may be useful.

For three-dimensional crystals, planes within the lattice may be defined by the Miller indices $(h'k'l')$. By analogy with a two-dimensional structure the lattice points may be arranged in rows classified by the indices $(h'k')$. The diffraction spots of a LEED pattern may be attributed to scattering at rows $(h'k')$ and therefore these indices can also be used to describe LEED patterns. The distance $d_{h'k'}$ between two neighbouring rows of direction $[h'k']$ is given by the relation

$$\frac{1}{d_{h'k'}^2} = \frac{h'^2}{b_1^2 \sin^2 \gamma} + \frac{k'^2}{b_2^2 \sin^2 \gamma} - \frac{2h'k' \cos \gamma}{b_1 b_2 \sin^2 \gamma} \tag{9.5}$$

where γ denotes the angle between the axes of the unit cell whose base vectors have lengths b_1 and b_2. If the unit cell is rectangular ($\gamma = 90°$) equ. (9.5) simplifies to

$$\frac{1}{d_{h'k'}^2} = \left(\frac{h'}{b_1}\right)^2 + \left(\frac{k'}{b_2}\right)^2. \tag{9.6}$$

9.3. Formation of the diffraction pattern

The basis for the interference of electrons at crystal surfaces is the de Broglie equation $\lambda = h/mv$. In the case of wave scattering at an array periodic in one dimension constructive interference takes place if the scattered waves from neighbouring lattice points have path differences of multiples of the wavelength λ. If the primary wave strikes the surface with an incident angle φ_0 interference of the backscattered waves occurs in directions φ (fig. 9.4), where φ is given by the condition

$$a (\sin \varphi - \sin \varphi_0) = n \lambda. \tag{9.7}$$

a is the distance between the periodically arranged scatterers and n an integer denoting the order of diffraction.

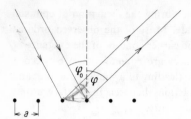

Fig. 9.4. Scattering of a plane wave at a one-dimensional periodic lattice.

An arrangement of lattice points which is periodic in two dimensions may be considered as an ensemble of parallel rows of scatterers with directions [h'k'] and mutual distances $d_{h'k'}$, as outlined in the foregoing section. In this case interference maxima are to be expected in directions φ given by

$$n\lambda = d_{h'k'} (\sin \varphi - \sin \varphi_0). \tag{9.8}$$

If, as is usually the case, LEED experiments are performed with normal incidence of the primary electrons ($\varphi_0 = 0$), then this equation simplifies to

$$\sin \varphi = \frac{n\lambda}{d_{h'k'}} \approx \frac{n}{d_{h'k'}} \cdot \sqrt{\frac{150}{U}} \tag{9.9}$$

with the electron energy U in [eV] and $d_{h'k'}$ in [Å] units. Each set of atomic rows $\{h'k'\}$ will therefore give rise to a series of diffraction maxima nh', nk' with varying order of diffraction n. Usually the order of diffraction n is included in the Miller indices h', k' giving the so-called Laue indices h $= n$h', k $= n$k'. Then the condition for interference at normal incidence becomes

$$\sin \varphi = \frac{1}{d_{hk}} \cdot \sqrt{\frac{150}{U}}. \tag{9.10}$$

Any particular reflex (hk) will appear for the first time at an electron energy $U = \dfrac{150}{d^2_{hk}}$ (i.e. $\varphi = 90°$ and $n = 1$). The larger the unit cell the closer the first diffraction maxima will be located to the surface normal. Increasing the electron energy leads to a movement of the diffraction maxima towards the (0,0)-reflex which corresponds to direct reflection.

The advantage of this simple treatment (which is analogous to the Bragg treatment of x-ray interferences in crystals) is that a complete description of the directions of the interference maxima is given, but the intensities remain completely unknown. The directions can be observed directly in a typical LEED apparatus, and they allow the d_{hk} to be determined and therefore the geometry of the unit cell. The arrangement of atoms within the unit cell (i.e. the basis) cannot be derived by analyzing the *positions* of diffraction maxima. As in the case of x-ray structure analysis this can only be achieved by analyzing the information contained in the *intensities*.

The need for this additional information is illustrated by fig. 9.5 where different possible arrangements of adsorbate atoms on a cubic substrate surface are shown. All these arrangements correspond to a c2 × 2 structure and lead therefore to identical diffraction patterns.

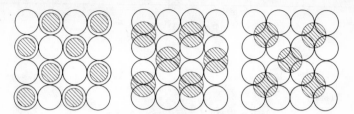

Fig. 9.5. Three different possible arrangements of adsorbed particles forming a $c\,2\times2$-structure.

9.4. Experimental equipment

9.4.1. Introduction

In standard x-ray structure analysis the information about the structure of the surface is contained both in the directions of the interference maxima, i. e. the geometry of the diffraction pattern, and in the intensities of the diffraction 'spots'. This is basically the same for LEED, and accordingly the experimental devices may be divided into systems which deliver the geometry of the diffraction pattern directly (but the intensities only indirectly) and into systems with which the intensities may be measured directly and afterwards be composed into the diffraction pattern, i. e. the allowed directions of diffraction maxima. It is also possible to combine both principles in a more complex apparatus. Most of the LEED systems currently in use are of the so-called 'display-type', where the complete diffraction pattern is registered but no direct measurement of the intensities is made. As yet the theory of LEED intensities is not sufficiently advanced to allow a routine structure analysis of a surface analogous to x-ray diffraction work. Therefore most of the investigations are restricted to the observation of diffraction patterns, but even this information alone is invaluable in surface research.
The essential elements of a LEED apparatus are

1. an electron gun for producing a reasonably parallel and monoenergetic electron beam with energies varying between ca. 20 and 500 eV.
2. a crystal holder with facilities for moving and heating the sample.
3. a detection system for the elastically scattered electrons.

9.4.2. Electron gun

The principle of construction of a typical LEED electron gun is contained in fig. 9.6. A directly or indirectly heated cathode within a Wehnelt cylinder emits electrons which are collimated by a lens system and finally leave the drift tube with the desired energy $e\,U$. This drift tube is usually at the same potential as the sample, so that the electrons traverse a field-free space to the crystal surface and are only deflected from their straight flight by stray magnetic fields. At lower electron energies it may be favourable to emit the electrons from the drift tube with a higher energy and then to decelerate them to the desired energy by applying a retarding potential to the crystal ("bias"). A typical electron beam has a diameter of 1 mm at the crystal

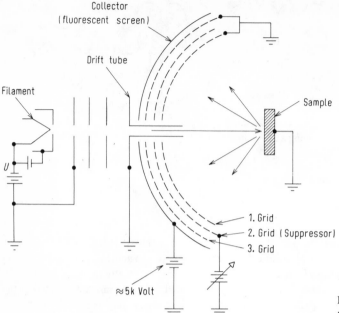

Fig. 9.6. Scheme of a LEED display-system.

surface. This is connected with a deviation in the parallelity of the electron beam. These limits are difficult to improve since they stem mainly from the finite size of the cathode. The situation may become even worse if space charges are built up, which is particularly likely at smaller electron energies.

The energy spread of the electrons is about 1 eV, caused mainly by the thermal energy distribution. The emission currents are usually of the order of 1 μA and vary with the electron energy, as the transmission of the gun changes. Recent electron guns may have electronic stabilisation to compensate for this.

Evaporation of cathode material can lead to contamination of the surface under investigation. This effect can be avoided by using an "off-axis-filament" device [23] where the beam is deflected by an additional electrode, which also has the advantage that weak patterns are not degraded by light from the cathode. Furthermore it is possible using this system to project a beam of light onto the sample through the electron optics.

9.4.3. Crystal manipulator

The single crystal sample must be mounted on a manipulator which provides facilities for different motions in ultra high vacuum; this is normally necessary for all of the surface techniques. A commercially available manipulator is shown in fig. 9.7, which allows rotation on the axis of the manipulator, as well as horizontal and vertical shifts. This design also allows rotation around an axis perpendicular to the surface plane which may become a necessary prerequisite for intensity measurements with varying diffraction geometry. Provision is also made for cooling the sample with liquid nitrogen and for indirect heating up to 1 200 °C. The

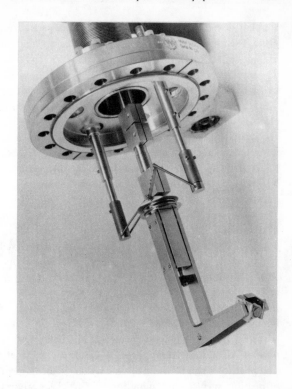

Fig. 9.7. A commercial crystal-manipulator (Varian).

minimum requirement is to be able to rotate the sample from analysis position to in front of the opening of the Argon ion gun in order to clean the surface. Many LEED systems are equipped with an extra electron gun for Auger electron spectroscopy which also makes an adjustment of the sample necessary.

The manipulator flange may also have several electrical feedthroughs mounted on it which can be used for heating the sample and measuring the temperature by means of a thermocouple. Sample heating is achieved by electron bombardment, indirectly by an oven, or most simply by direct resistance heating. Magnetic fields accompanying high electric currents may disturb the LEED observations. This effect can be avoided by the use of the so-called "grid-blanking" technique using rectified alternating current [24]. During one half-cycle the crystal is heated and the formation of the LEED pattern is suppressed; and during the next half cycle the LEED pattern is observed. Studies of physical adsorption etc. may need cooling of the sample, which can be achieved by special manipulators [25]. Facilities for fracturing crystals in ultrahigh vacuum have also been constructed.

9.4.4. Detector system

Most of the LEED systems presently in use and all types which are commercially available are of the "display-type" (fig. 9.6): Electrons leave the drift tube of the electron gun with the desired energy U and strike the sample surface. Normal incidence is usually employed. In the simplest device the back-scattered electrons pass a system of two hemispherical concentric grids. The first grid is grounded as are the crystal and the drift-tube so that scattered electrons are not deflected electrostatically in the field free region between sample and the first grid.

The addition of Helmholtz coils around the LEED chamber may be necessary in order to avoid magnetic disturbances e. g. by the earth's field.

The second grid is at a negative potential whose magnitude is slightly smaller than the primary electron energy U which therefore repels the inelastically scattered electrons. After passing the second grid the elastically scattered electrons are accelerated onto a (also concentric) fluorescent screen by a positive potential of a few kilovolts. The screen then will exhibit "diffraction spots" at the positions of the interference maxima. The diffraction pattern may be observed visually through a window in the vacuum chamber. A permanent record is made by photographing the display.

The arrangement using only two grids causes considerable field inhomogeneities in the meshes of the second grid under the influence of the high voltage of the fluorescent screen. A marked improvement of the LEED patterns therefore is achieved by the addition of a third grid at the same potential as the second (repeller) grid [26]. Many LEED systems are further equipped with a fourth grid which improves the resolution properties of the grid system when it is used as an energy analyzer for Auger spectroscopy [27].

Instead of the spherical symmetry of the grids and the fluorescent screen another possibility is a cylindrical arrangement [28].

A further method has been developed by Tucker [20] on the basis of a previous proposal by Sproull [11] in which a magnetic field is directed perpendicular to the primary electron beam. The primary beam is thus deflected by 90° before striking the surface, diminishing the danger of possible surface contamination by particles evaporated from the cathode. The diffracted electrons are also deflected, which also serves to filter off the inelastically scattered electrons so that no retarding grid is necessary. A disadvantage of this arrangement is that the diffraction pattern is heavily distorted which is not the case with the usual display technique.

The second class of detector systems is based on the initial experimental devices of Davisson and Germer [6] and Farnsworth [29], and is shown schematically in fig. 9.8. The intensity

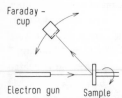

Faraday – cup

Electron gun   Sample   Fig. 9.8. Principle of the Faraday-cup LEED systems

of the back-scattered electrons is measured by a movable Faraday cup, which again filters out the inelastically scattered electrons by a retarding potential between electrodes at the orifice. An improved performance can be obtained using a lens system instead of grids which avoids secondary emission. The crystal may be rotated around an axis in the direction of the primary electron beam which enables the registration of all beams with equal diffraction angles. The angle of diffraction may be varied by changing the position of the Faraday cup, which allows each position on the reciprocal lattice to be reached within certain limits. This method enables very accurate measurements of the LEED intensities to be made, but it is very tedious, since it needs many hours to investigate a complete diffraction pattern thus preventing the observation of fast changes in the surface structure.

This disadvantage is eliminated to a great extent by an automatic system designed by Park and Farnsworth [30]. In a device similar to that shown in fig. 9.8 the crystal is rotated very quickly and simultaneously the position of the collector is altered by a slower movement, so that the orifice of the Faraday cup moves on a spiral across the reciprocal lattice. The registered intensities control the emission of the cathode of a storage oscilloscope whose potentials across the x- and y-deflection plates are varied in step with the movement of the crystal and the collector. It is thus possible to display a complete diffraction pattern on the screen of an oscilloscope within a few minutes. In addition it is also possible to measure intensity versus voltage (I/U) curves with high accuracy e. g. for a fixed position of the collector. Fig. 9.9 shows

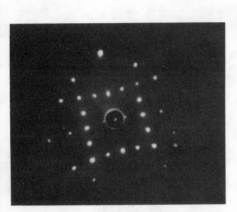 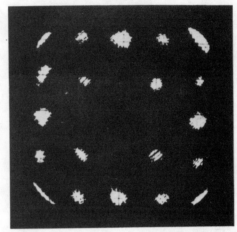

Fig. 9.9. Diffraction patterns from a c(4×2)/$45°$-structure caused by CO adsorbed on Pd(100), using a) the normal display technique [31], b) the automatic Faraday-cup system [19].

two diffraction patterns from CO covered Pd(100) surfaces, which were obtained by the Park-Farnsworth method [19] and by the usual display-technique [31].

An alternative possibility which combines the advantages of the display-type technique and the Faraday-cup method consists of a combination of both detector systems. Fig. 9.10 shows a commercial system of this type which enables a Faraday cup to be positioned over any diffraction spot on the fluorescent screen in order to measure its intensity. The two independent motions necessary are controlled from outside the vacuum.

If attempts are made to compare experimental I/U-curves with theoretical calculations it would be desirable to measure only the intensities in the centers of the diffraction spots in order to minimize the contributions from thermal diffuse scattering. In all practical cases however the measured intensities represent an integration over a more or less large region of the cross-section of the beam, usually $\sim 1°$. If absolute intensities are desired it is of course also necessary to measure the primary current i_0. For relative I/U-curves i_0 has best to be held constant (i. e. independent of the energy) which may be achieved electronically in some systems. An alternative method for evaluating intensities consists in measuring the brightness of diffraction spots by a spot photometer.

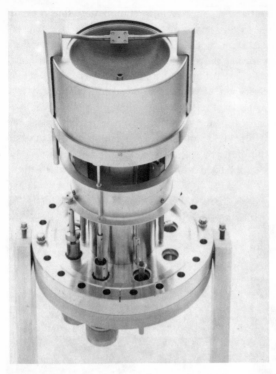

Fig. 9.10. A commercial LEED system consisting of a four-grid display optics and a movable Faraday-cup (Varian).

9.5. Geometrical theory of diffraction

9.5.1. Introduction

In section 9.3 the formation of a diffraction pattern was described on the basis of phase shifts by multiples of the wavelength λ between the back-scattered waves originating from neighbouring lattice points. The resulting Bragg equation determines the directions of the interference maxima (Bragg beams) and their positions on the fluorescent screen, but does not contain any information about the intensities. This approach forms the basis of the geometrical theory of diffraction which provides the geometry of the unit cell without including the LEED intensities. This approach is justified for x-ray analysis from the general scattering theory through the intensities and the Laue equations, from which the validity of the Bragg equation is easily proved [32].

In principle the diffraction of slow electrons at crystal surfaces must be treated from a similar starting point, namely on the basis of scattered intensities. As becomes clear later the so-called kinematic diffraction theory is a very good approach for the description of x-ray interference phenomena but has only very limited validity for the LEED problem. Owing to the strong interaction between low energy electrons and matter it must be replaced by a much more complicated 'dynamic' theory. This dynamic theory has not yet reached a stage which enables the straight forward analysis of real surface structures (with a few remarkable exceptions!) This handicap restricts the amount of information which can be obtained from LEED measurements at present. However it can be shown that the validity of the geometrical theory and the information so derived for the positions of the diffraction maxima is not influenced by these complications, although in certain cases it is necessary to consider the formation of additional diffraction spots caused by multiple scattering. In surface chemistry therefore most LEED investigations ignore the intensity information and are confined to an analysis of the diffraction pattern based on the simple geometrical theory. Frequently workers restrict themselves to describing only the diffraction pattern, but even so unique conclusions about surface processes can be drawn. In favourable cases it is possible to deduce plausible models of surface structures only from the knowledge of the unit cell (derived from the geometric theory) with the help of information about the system under investigation obtained from other sources.

In simple cases it is no problem for an experienced worker to deduce the unit cell of the surface structure simply by inspecting the diffraction pattern. In more complicated cases it is convenient to use a general concept based on the construction of the reciprocal lattice.

9.5.2. The reciprocal lattice

The elastic interaction of electrons with a surface is treated as the scattering of waves at a two-dimensional lattice, which means that the finite thickness of the surface region and penetration of the electrons to deeper layers is neglected. This two dimensional lattice represents the simplest arrangement of points with a periodicity equal to that of the atoms in the surface layer. The unit cell with the basis vectors a_1, a_2 is the smallest parallelogram from which the lattice may be constructed by translation operations.

A complementary reciprocal lattice is then defined by the basis vectors a_1^*, a_2^* obeying the equations

$$a_i \cdot a_j^* = \delta_{ij} \qquad (i, j = 1, 2) \tag{9.11}$$

The Kronecker symbol $\delta_{ij} = 0$ if $i \neq j$, and $\delta_{ij} = 1$ if $i = j$. In other words, $a_1^* \perp a_2$, and $a_2^* \perp a_1$. The reciprocal lattice for the two-dimensional case can be derived easily with the methods used for three-dimensional lattices commonly in use for x-ray analysis [33].

The above definition leads to the following rules for constructing the two-dimensional reciprocal lattice from a given real lattice (fig. 9.11): a_1^* is perpendicular to a_2, and the length a_1^* is derived from the condition $a_1 \cdot a_1^* = 1$. Hence $a_1 \cdot a_1^* = 1/\cos\left(\frac{\pi}{2} - \gamma\right)$ where γ is the angle between a_1 and a_2, or

$$a_1^* = 1/(a_1 \sin \gamma). \tag{9.12}$$

By analogy $a_2^* = 1/(a_2 \sin \gamma)$, where $a_2^* \perp a_1$.

The angle γ^* between a_2^* and a_2^* is given by $\gamma^* = \gamma + 2\left(\frac{\pi}{2} - \gamma\right) = \pi - \gamma$, which means $\sin \gamma^* = \sin \gamma$. The area A of the elementary cell of the real lattice is given by

$$A = a_1 a_2 \sin \gamma = |a_1 \times a_2|. \tag{9.13}$$

The corresponding relation for the reciprocal lattice is

$$A^* = a_1^* a_2^* \sin \gamma^* = |a_1^* \times a_2^*| = 1/A. \tag{9.14}$$

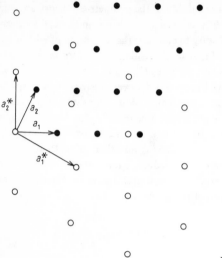

Fig. 9.11. A two-dimensional real lattice, described by a_1, a_2 (dark circles), and its reciprocal lattice a_1^*, a_2^* (open circles).

If a surface structure with basis vectors b_1 and b_2 superimposed on a substrate lattice a_1, a_2 is present then as shown in section 2 both lattices a and b may be related by $b = \mathfrak{M}a$ or

$$b_1 = m_{11} a_1 + m_{12} a_2$$

$$b_2 = m_{21} a_1 + m_{22} a_2. \tag{9.2}$$

The reciprocal lattice b^* may be correlated with the reciprocal lattice a^* in a similar manner:

$$b_1^* = m_{11}^* a_1^* + m_{12}^* a_2^*$$

$$b_2^* = m_{21}^* a_1^* + m_{22}^* a_2^*. \tag{9.15}$$

It will be shown that the m_{ij}^* may be obtained from an inspection of the diffraction pattern. The remaining task consists of the evaluation of the unknown factors m_{ij} needed to derive the unit cell of the surface structure in terms of the substrate lattice.

It can be shown that $\mathfrak{M}^*$ is the inverse transposed matrix of $\mathfrak{M}$, so that $\mathfrak{M}^* = \tilde{\mathfrak{M}}^{-1}$ and $\mathfrak{M} = \tilde{\mathfrak{M}}^{*-1}$, which leads to the following relations:

$$m_{11} = \frac{1}{\det \mathfrak{M}^*} \cdot m_{22}^*$$

$$m_{12} = -\frac{1}{\det \mathfrak{M}^*} \cdot m_{21}^* \tag{9.16}$$

$$m_{21} = -\frac{1}{\det \mathfrak{M}^*} \cdot m_{12}^*$$

$$m_{22} = \frac{1}{\det \mathfrak{M}^*} \cdot m_{11}^*$$

where $\det \mathfrak{M}^* = m_{11}^* \cdot m_{22}^* - m_{21}^* \cdot m_{12}^*$.

These equations allow the real surface lattice to be derived simply from the corresponding reciprocal lattice.

9.5.3. Interference conditions and the Ewald construction

The principle underlying the LEED experiment is drawn schematically in fig. 9.12.

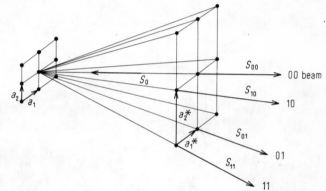

Fig. 9.12. The principle of the formation of a diffraction pattern in a LEED experiment. The size of the unit cell a_1, a_2 of the real lattice is of course strongly magnified in this drawing.

The direction of the impinging primary beam is characterized by the unit vector s_0, and the directions of interference maxima are given by a set of unit vectors s, which are determined by the following Laue conditions

$$a_1 \cdot (s - s_0) = h_1 \lambda$$

$$a_2 \cdot (s - s_0) = h_2 \lambda \tag{9.17}$$

(h_1, h_2 = integers). These equations follow from the theory for the scattered intensity (see section 9.6.2). For the case of a one-dimensional periodic arrangement it is immediately evident from fig. 9.12 that these relations are identical with the originally derived equation (9.7), i.e. the Laue conditions are equivalent to the Bragg relation.

The problem of deriving the directions s from the above equations is solved by the expression

$$\frac{\Delta s}{\lambda} = h_1 a_1^* + h_2 a_2^* \tag{9.18}$$

where

$$\Delta s = s - s_0. \tag{9.19}$$

This can be proved by insertion in equ. (9.17) and using the definition of the reciprocal lattice:

$$h_1 \lambda = \boldsymbol{a}_1 \cdot \Delta s \overset{?}{=} \lambda \boldsymbol{a}_1 \cdot (h_1 \boldsymbol{a}_1^* + h_2 \boldsymbol{a}_2^*) \overset{!}{=} h_1 \lambda$$

$$h_2 \lambda = \boldsymbol{a}_2 \cdot \Delta s \overset{?}{=} \lambda \boldsymbol{a}_2 \cdot (h_1 \boldsymbol{a}_1^* + h_2 \boldsymbol{a}_2^*) \overset{!}{=} h_2 \lambda$$

which shows that the directions of the interference maxima (i.e. of the scattered beams) are determined by the vectors of the reciprocal lattice

$$\boldsymbol{g} = h_1 \boldsymbol{a}_1^* + h_2 \boldsymbol{a}_2^*. \tag{9.20}$$

It is possible to determine the scattered beams simply by using the so-called Ewald construction (fig. 9.13). A set of lines perpendicular to the crystal surface is drawn through the points of the

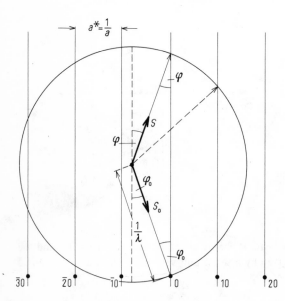

Fig. 9.13. The Ewald construction for a one-dimensional lattice.

reciprocal lattice. The Ewald sphere is now constructed with a radius of length of $1/\lambda$ and, with the centre at a point at $-\frac{1}{\lambda} \cdot \boldsymbol{s}_0$ from the origin of the reciprocal lattice*). The points where the sphere intersects the reciprocal lattice rods determine the directions $\boldsymbol{s}$ of the scattered beams. The validity of this statement is obvious for the one-dimensional case drawn in fig. 9.13, since

$$\frac{1}{\lambda} \cdot \sin \varphi = \frac{1}{\lambda} \cdot \sin \varphi_0 + h_1 \cdot \frac{1}{a} \tag{9.7}$$

or

$$a(\sin \varphi - \sin \varphi_0) = h_1 \lambda.$$

---

*) Sometimes in the literature the length of the radius of the Ewald sphere is taken as $|\boldsymbol{k}_0| = \frac{2\pi}{\lambda}$ instead of $1/\lambda$. The reciprocal lattice vectors $\boldsymbol{g} = h_1 \boldsymbol{a}_1^* + h_2 \boldsymbol{a}_2^*$ are then to be replaced by a set $\boldsymbol{g}' = 2\pi(h_1 \boldsymbol{a}_1^* + h_2 \boldsymbol{a}_2^*)$.

In LEED experiments the directions s of the scattered beams are given by the points where they intersect with the fluorescent screen giving rise to the "diffraction spots", as shown in fig. 9.12. As a consequence, the *observed diffraction pattern is a direct representation of the reciprocal lattice of the surface*.

In typical cases $1/\lambda$ is of the same order of magnitude as $1/a$. This implies that only a few diffracted beams, i.e. diffraction spots, will appear. Increasing $1/\lambda$ (i.e. increasing the electron energy) leads to an increase in the radius of the Ewald sphere, which means that more diffraction spots are possible which move in continuously towards the (0,0)-spot with increasing energy. The (0,0)-spot is caused by direct reflection of the primary beam at the surface without diffraction. It does not change its position in the diffraction pattern with varying electron energy and is therefore easily identified.

9.5.4. Analysis of a simple diffraction pattern

How the unit cell of a surface structure may be determined in practice is illustrated in the following example of a structure on the (110) surface of a fcc crystal.

A sketch of the diffraction pattern (and therefore also the reciprocal lattice) is shown in fig. 9.14a.

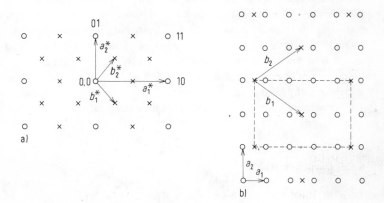

Fig. 9.14. Analysis of a simple diffraction pattern.
a) Reciprocal lattice ($=$LEED pattern), o: Substrate ('normal') spots, $\times$: Overlayer ('extra') spots, b) Real lattice of the substrate (o) and the substrate ($\times$). The dashed line denotes the non-primitive $c\,4\times 2$-unit cell.

The 'extra spots' caused by the adsorbed surface structure are marked differently.

The substrate diffraction spots have been indexed using the common nomenclature which automatically yields the reciprocal substrate unit vectors a_1^* and a_2^*, where the shorter axis in the real lattice (i.e. the longer in the reciprocal lattice) is indexed by 1 ($\hat{=}h_1$). Every point on the reciprocal lattice of the surface structure can be reached by translation operations in terms of the basis vectors b_1^* and b_2^*. These are interrelated with the reciprocal unit vectors of the substrate by the relations

$$b_1^* = \tfrac{1}{4} a_1^* - \tfrac{1}{2} a_2^* = m_{11}^* a_1^* + m_{12}^* a_2^*$$

$$(9.21)$$

$$b_2^* = \tfrac{1}{4} a_1^* + \tfrac{1}{2} a_2^* = m_{21}^* a_1^* + m_{22}^* a_2^*$$

(Of course, other choices for b_1^*, b_2^* are possible without influencing the final result, for example $b_1^* = \tfrac{1}{2} a_1^*$; $b_2 = \tfrac{1}{4} a_1^* + \tfrac{1}{2} a_2^*$).

Next, the substrate lattice is constructed from the position of the substrate diffraction spots (this is usually already known): a_1 is perpendicular to a_2^* and has the length $a_1 = \dfrac{1}{a_1^* \sin \gamma^*} = \dfrac{1}{a_1^*}$.

Similarly $a_2 \perp a_1^*$ with $a_2 = \dfrac{1}{a_2^*} = \sqrt{2}/a_1^*$ (fig. 9.14b).

The real lattice of the surface structure $b_1 = m_{11} a_1 + m_{12} a_2$; $b_2 = m_{21} a_1 + m_{22} a_2$ is derived using the relations (9.16) yielding $b_1 = 2a_1 - a_2$; $b_2 = 2a_1 + a_2$. This unit cell and the real lattice of the surface structure thereby defined are also drawn in fig. 9.14b. It can be seen that in this case another (larger) unit cell with the symmetry of the substrate, which is not primitive but centered, is also possible and which is called a $c4 \times 2$-structure in Wood's nomenclature.

With some experience this result can be deduced from the diffraction pattern by inspection. Every second spot of a "true" (i.e. primitive) 4×2-structure is missing which is caused by a second identical scatterer in the centre of the 4×2-unit cell. Characteristic extinctions of diffraction spots are one of the subjects of the treatment of diffracted intensities.

Since the diffraction pattern (i.e. the positions of the diffraction spots) is determined only by the geometry of the surface structure, any shifts of the surface atoms parallel to b_1 and b_2 with respect to the topmost atoms of the substrate lead to identical positions of 'extra spots'. It is therefore not possible to derive the position of adsorbed particles relative to the substrate atoms from such observations. Furthermore surface lattice points composed of groups of atoms cannot be ruled out, but if the unit cell of the surface structure is small such possibilities may be precluded since the adsorbed particles have finite dimensions. In these cases the unit vectors of the surface structure determine the configurations of adsorbed particles with respect to each other.

9.5.5. Domain structures

Even the diffraction patterns of simple structures may contain a series of complications, such as the existence of domains which means that the periodic sequence of elementary cells b_1, b_2 of the surface structure may not cover the whole surface area (i.e. the area covered by the primary beam), but may be discontinuous. (This effect should not be confused with the existence of point defects which are always present in an otherwise perfect periodic arrangement.) The influence on the diffraction pattern depends on the mean diameter d of the regions with uniform periodicity as compared with the coherence width L of the electron radiation. The latter is controlled by experimental considerations and is discussed in section 9.7.2 in more detail. L is typically in the range of $50-100$ Å. If $d < L$ the diffraction pattern contains effects due to the superposition of the *amplitudes* of the electron waves scattered at the different regions leading to effects such as spot broadening or spot splitting, formation of streaks, etc. which will be treated in section 9.7.

If $d > L$ the diffraction pattern consists of a superposition of the *intensities* of the interference phenomena caused by the separate regions with uniform periodicities, i.e. a superposition of the different sets of diffraction spots, since the primary electron beam (diameter ~ 1 mm) usually covers many such domains.

However, it is also possible that the size of these domains reaches macroscopic dimensions, so that by shifting the electron beam across the surface different domains may be analyzed separately. The preferential formation of a certain type of domain may sometimes be initiated artificially by preparing a surface whose orientation deviates by a few degrees from the nominal low index crystallographic orientation, with the result that one type of domain of an adsorbate structure is formed predominantly, leading to the abolition of the symmetric distribution of spot intensities in the diffraction pattern [35]. Similar effects have been observed with semiconductor surfaces prepared by cleavage [36].

If we assume the existence of two types of domains α and β the following situations may occur:

a) The basis vectors associated with α and β include different angles and/or have different lengths: $\measuredangle({}^{\alpha}\boldsymbol{b}_1, {}^{\alpha}\boldsymbol{b}_2) \neq \measuredangle({}^{\beta}\boldsymbol{b}_2)$ and/or $|{}^{\alpha}\boldsymbol{b}_1| \neq |{}^{\beta}\boldsymbol{b}_1|, |{}^{\alpha}\boldsymbol{b}_2| \neq |{}^{\beta}\boldsymbol{b}_2|$.
This means that the diffraction patterns corresponding to α and β cannot be made coincident by rotation or mirror plane operations. These cases are usually not very complicated and may

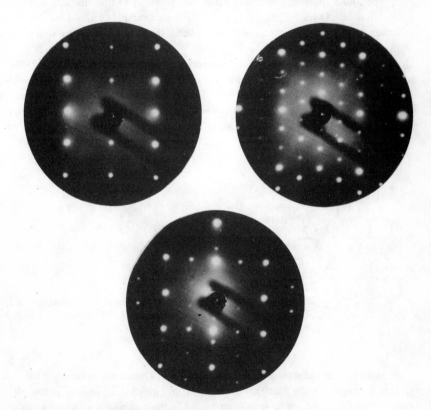

Fig. 9.15. Diffraction patterns for oxygen adsorbed on Cu(110).
a) 2×1-structure at low coverage, b) $c6 \times 2$-structure at high coverage, c) Superposition of diffraction patterns from both structures at intermediate coverages indicating the coexistence of domains of the 2×1- and $c6 \times 2$-structures. After [56].

be easily identified as superimposed diffraction patterns. This is observed for example if adsorbed phases with different structures coexist on the surface.

Oxygen forms a 2×1-structure on Cu(110) at low exposure and at higher exposures a c6 × 2-structure [37]. The transition is characterized by a superposition of both systems of diffraction spots (fig. 9.15) indicating that adsorbed oxygen exists in the form of sizeable domains with one or other configuration. The variation of the total surface concentration is accompanied by a continuous increase of the intensities of one set of diffraction spots and by a simultaneous continuous decrease of the intensities of the second system of 'extra' spots due to the other type of domain. Similar situations are quite common if two different kinds of particles are adsorbed ('competitive adsorption').

b) The elementary cells of α and β may be made identical by rotation or mirror plane operations. This situation will always occur in cases where the unit cell of a surface structure has a lower symmetry than that of the substrate. For example a rectangular unit cell on a square substrate lattice will have two completely equivalent positions rotated by 90° to each other. Therefore the diffraction pattern will be composed from two sets of diffraction spots rotated by 90° to each other, both together presenting again the symmetry of the substrate diffraction pattern. Adsorption of CO on Pd(100) [19] (fig. 9.9) is an example of this kind where the domain structure does not follow immediately from the inspection of the diffraction pattern but only from a more detailed analysis.

However there are also cases where it is in principle impossible to decide from the diffraction pattern whether this is caused by a domain structure or by a uniform structure with a different unit cell with higher symmetry. The diffraction pattern caused by oxygen adsorption on Pd(111) [38] is an example of this type and it is not clear whether this is due to a 2×2-structure or to three domains of a 2×1-structure with different orientations as shown in fig. 9.16.

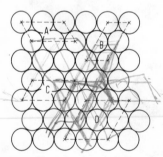

Fig. 9.16. Three types of domains of a 2×1-structure on a (111) surface (A, B, C) will cause the same LEED pattern as a 2×2-structure (D).

c) α and β are identical after a translation operation. This situation occurs if the unit cell of the surface structure is larger than that of the substrate, which is the case for practically all ordered surface structures! As can be seen from fig. 9.17a the origin of the surface structure may then be shifted into other equivalent positions. A somewhat different example of the same type occurs if the coordination of an adsorbed particle has a lower symmetry than that of the topmost atoms of the substrate, as illustrated in fig. 9.17b. Since translations of the unit cell of a

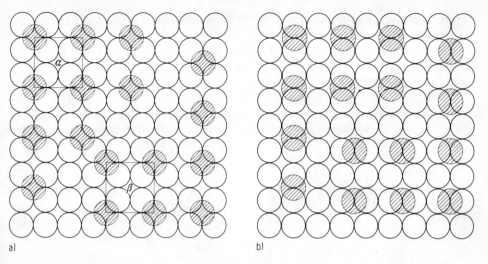

Fig. 9.17. Two types of domain structures which become identical by translation operations of their unit cells.

surface structure with respect to the substrate lattice do not alter the diffraction pattern the existence of such domains cannot be detected, unless the domain diameter becomes smaller than the coherence width, which will be discussed in section 9.7.

A detailed discussion of the numbers of equivalent domains which may exist under given conditions has been performed by May [17g]. The fundamental importance of the domain problem in surface crystallography is quite apparent and has a direct bearing on all theoretical attempts to calculate LEED intensities.

9.5.6. LEED patterns of coincidence lattices and incoherent structures

So far the scattering of electrons has been treated separately for two-dimensional lattices of the substrate and surface structures, and the diffraction pattern was then composed of both contributions which yields the positions of all diffraction spots correctly only for simple surface structures. The basis of such a procedure lies in the assumption that only single-scattering phenomena of the primary electrons occur – an approach which leads to the kinematic theory for the intensities which is well justified for x-ray diffraction. Since slow electrons interact very strongly with matter this picture is certainly not correct in the case of LEED, and effects due to multiple scattering must also be considered. For example it is possible that a wave which has been scattered at the substrate lattice will be scattered again at the lattice of the surface structure (which has a different periodicity). In the case of coincidence lattices and incoherent structures these multiple scattering effects will produce new additional diffraction spots which would not occur from a simple superposition of the diffraction patterns from the separate lattices. For a surface structure described by the reciprocal lattice vectors $^1g = {}^1h_1 \boldsymbol{b}_1^* + {}^1h_2 \boldsymbol{b}_2^*$ on top of a substrate lattice with $^2g = {}^2h_1 \boldsymbol{a}_1^* + {}^2h_2 \boldsymbol{a}_2^*$ the diffraction pattern contains only spots at positions given by ^1g and ^2g provided that only single-scattering occurs. This case is illustrated in fig. 9.18a for a one-dimensional example of a coincidence lattice where $3b = 4a$.

However multiple scattering leads to the occurrence of an effective reciprocal lattice [39] given by

$$g = {}^1h_1 \, a_1^* + {}^1h_2 \, a_1^* + {}^2h_1 \, b_1^* + {}^2h_2 \, b_2^* \tag{9.22}$$

so that the effective reciprocal lattice which is equivalent to the diffraction pattern contains the periodicities of the reciprocal substrate and surface lattices and *their linear combinations*.

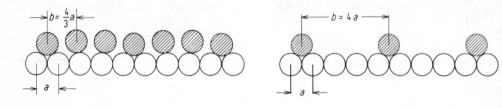

Diffraction patterns

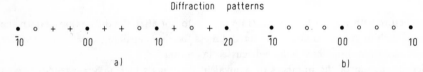

Fig. 9.18. A coincidence lattice a) and a true superstructure b) cannot be distinguished from their LEED patterns.
● 'normal' substrate spots, ○ : single diffraction overlayer sports, + : multiple diffraction overlayer spots

In order to find the cases where this effect causes additional diffraction spots (sometimes also called "satellite spots" [40]) again the correlation between both lattices is considered. From

$$b_1 = m_{11} \, a_1 + m_{12} \, a_2$$
$$b_2 = m_{21} \, a_1 + m_{22} \, a_2 \tag{9.2}$$

by means of simple operations the following relations for the reciprocal lattices may be derived using the equations (9.16):

$$a_1^* = m_{11} \, b_1^* + m_{21} \, b_2^*$$
$$a_2^* = m_{12} \, b_1^* + m_{22} \, b_2^* \tag{9.23}$$

Simple structures are defined by $m_{ij} =$ integers. Then the reciprocal lattice vectors a_1^*, a_2^* of the substrate are, according to equ. (9.23), part of the reciprocal lattice of the surface structure $^2g = {}^2h_1 \, b_1^* + {}^2h_2 \, b_2^*$, because $^2h_1$ and $^2h_2$ are any integers. Consequently the superposition of

^1g and ^2g always creates vectors already present in the set of ^2g. With simple structures, therefore, no additional spots due to multiple scattering effects are created.

However the situation is different for coincidence lattices and incoherent structures, where not all m_{ij}'s are integers and therefore ^1g is not included in ^2g. A relatively simple example of this type [41] is shown in fig 9.19. This is a diffraction pattern of a hexagonal structure caused

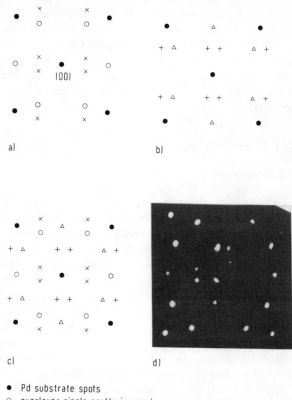

Fig. 9.19. Analysis of the diffraction pattern caused by Xe adsorption on Pd(100). The structure model is reproduced in fig 11.7. a) Diffraction pattern from the first domain orientation with single and multiple diffraction spots. b) Pattern from the second domain orientation. c) Superposition of a) and b). d) Observed LEED pattern. After Palmberg [41].

● Pd substrate spots
○ overlayer single scattering spots
× overlayer multiple scattering spots } first domain orientation
△ overlayer single scattering spots
+ overlayer multiple scattering spots } second domain orientation

by adsorption of Xe on Pd(100). (The details are given in 11.4.) As a further complication two different domain orientations have to be taken into account. It is clear that the analysis of a diffraction pattern may be complicated considerably by such effects.

In particular it is impossible just using the LEED pattern to make a distinction between a "true" superstructure with large lattice constants (e.g. $b = 4a$ in fig. 9.18b) and a coincidence lattice with identical combined periodicity (superimposed from surface and substrate lattice) (e.g. $3b = 4a$ in fig. 9.18a) [39]. It has been assumed previously that such a decision could be made by observing the disappearance of the "extra" spots at low electron energies, but this is not justified [39, 42].

The possible positions of "satellite" spots in the diffraction pattern for a given surface and substrate structure may be constructed simply from equ. (9.22). It has been observed however that the intensities of these spots may vary considerably.

It is necessary to point out that it is also possible to interpret the "satellite" spots without the concept of multiple scattering by using an extension of the kinematic (single scattering) picture. As can be seen in fig. 9.18a the substrate atoms are "shadowed" by the atoms of the adsorbed layer to various extents. Therefore the scattering factors of the latter are modulated periodically leading to the appearance of "forbidden" diffraction spots [17a, g]. Another possibility is that of a periodic variation of the positions of the topmost substrate atoms [43, 44]. A detailed discussion of these effects has been published by Palmberg and Rhodin [43], who suggest that the effect of displacements of the atoms should dominate at higher electron energies, whereas effects due to multiple scattering should determine the relative spot intensities at very low energies. Of course these problems can only be settled finally by a complete theory for LEED intensities.

9.6. Kinematic Theory

9.6.1. Introduction

The kinematic theory of scattered intensities is based on the assumption that the incident radiation interacts with matter only very weakly so that only single scattering processes have to be considered to give a good approximation. This assumption is justified for the scattering of x-rays and fast electrons, but does not hold for the interaction of low energy electrons with solids. Direct evidence of multiple scattering effects has already been discussed in connection with the analysis of diffraction patterns and is of even greater importance in the treatment of spot intensities as a function of energy (I/U curves). A complete analysis of surface structures cannot be expected simply from the kinematic scattering theory. Nevertheless nearly all problems of "real" surface crystallography have been treated using this approximation for the following reasons: a) The theory is rather simple. b) The results are correct as far as the positions of diffraction spots are concerned. This enables an extension of the geometrical theory to partially disordered structures. c) The theory can provide useful results in cases where only relative intensities are needed, but due caution must be exercised. d) An attempt can be made to extract the "kinematic" part contained in the intensity data. If this can be achieved a structural analysis by means of the kinematic theory becomes possible. This is the basic idea of the so-called "averaging" methods which may become of some significance in surface crystallography.

9.6.2. Scattering at two-dimensional lattices

The scattering of an electron wave at an isolated atom will be considered first. The incident beam of electrons is represented by a plane wave with wavelength λ moving in the direction s_0, where s_0 is a unit vector. This wave may be described by

$$\psi^0 = \psi_0 \cdot e^{i k_0 \cdot r} \qquad (9.24)$$

where $k_0 = \dfrac{2\pi}{\lambda} \cdot s_0$ is the wave vector (which represents also the electron momentum) [45].

At a point of observation R which is assumed to be at a large distance from the scatterer (fig. 9.20) a scattered wave arrives described by the wave vector $k = \dfrac{2\pi}{\lambda} \cdot s$. If this wave originates from scattering at an atom j at the position R_j its amplitude is given by

$$\psi = \left(\psi_0 \cdot \frac{e^{ik \cdot R}}{R} \right) \cdot f_j(k_0, k)\, e^{i(k-k_0) \cdot R_j} \tag{9.25}$$

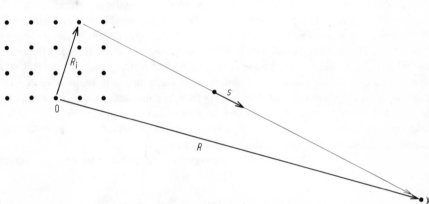

Fig. 9.20. Scattering at the points of a lattice. X = point of observation.

The first term describes a spherical wave. f_j is the atomic scattering factor which for a given kind of atom depends on k_0 and k and therefore on the wavelength and the directions of the incident and scattered waves. The last term denotes the phase shift between a wave scattered at R_j and the origin of the coordinate system caused by the path difference.

If a whole ensemble of atoms is considered then the total scattered wave results from a super-position of the waves originating from the single atoms with the corresponding phase shifts:

$$\psi \propto \sum_j f_j(k_0, k)\, e^{i(k-k_0) \cdot R_j} \tag{9.26}$$

If scattering takes place at a two-dimensional periodic lattice the atomic scattering factors f_j may be replaced by the scattering factor F (the so-called structure factor) of the unit cell, which of course has to be identical for all cells due to the periodicity of the lattice. The structure factor F depends on the atomic scattering factors f_j and on the positions of the atoms within the unit cell. If the two-dimensional lattice is composed of $M_1 \cdot M_2$ lattice points, the unit cell being defined by vectors a_1, a_2, then $R_i = n_1 a_1 + n_2 a_2$, where $1 \leqslant n_1 \leqslant M_1$, $1 \leqslant n_2 \leqslant M_2$, and R_i reaches each lattice point on the surface. Equ. (9.26) may then be transformed into

$$\psi \propto F \cdot \sum_{n_1=1}^{M_1} e^{i n_1 a_1 \cdot (k-k_0)} \cdot \sum_{n_2=1}^{M_2} e^{i n_2 a_2 \cdot (k-k_0)} \tag{9.27}$$

or

$$\psi \propto F \cdot G \tag{9.28}$$

The so-called lattice factor G is an abbreviation of the product of the two sums, and is determined by the periodicity a_1, a_2 of the surface and by $(k - k_0)$ (i.e. the angle between the incident and the scattered beam).

The measurable quantity is not the amplitude but the intensity I of diffraction in a given direction which may be defined as the ratio of the electron current in the diffracted beam i to that in the primary beam i_0. The intensity is given by the square of the amplitude:

$$I \propto |F|^2 \cdot |G|^2 \tag{9.29}$$

This relation is of general validity for two-dimensional periodic structures, even within the framework of a complete dynamical theory. The difference between the kinematical and a dynamical theory is contained in the structure factor F. According to the single-scattering picture F is composed from a superposition of the atomic scattering factors multiplied by the corresponding phase shifts caused by the positions of the individual atoms within the unit cell, whereas a dynamic theory includes all the effects of multiple scattering, inelastic interaction etc. in the calculation of the scattered wave originating from the unit cell. In general F depends both on k as well as on k_0 separately, whereas the kinematical part of the structure factor is only a function of the difference $(k - k_0)$. This effect is of fundamental importance for one of the averaging techniques [46] [47] as discussed in section 9.11.

The so-called interference function $J = |G|^2$ may be evaluated easily by performing the summations [45]. The result is

$$J = \frac{\sin^2\left[\frac{1}{2} M_1 a_1 \cdot (k - k_0)\right]}{\sin^2\left[\frac{1}{2} a_1 \cdot (k - k_0)\right]} \cdot \frac{\sin^2\left[\frac{1}{2} M_2 a_2 \cdot (k - k_0)\right]}{\sin^2\left[\frac{1}{2} a_2 \cdot (k - k_0)\right]} \tag{9.30}$$

Maxima in the intensity are to be expected for directions k for which the arguments of *both* factors in the denominator are multiples of π, that means if

$$\tfrac{1}{2} a_1 \cdot (k - k_0) = h_1 \pi \qquad \text{and} \qquad \tfrac{1}{2} a_2 \cdot (k - k_0) = h_2 \pi \tag{9.31}$$

$(h_1, h_2 = \text{integers})$;

or, since $k = \dfrac{2\pi}{\lambda} \cdot s$,

$$a_1 \cdot (s - s_0) = h_1 \lambda \qquad \text{and} \qquad a_2 \cdot (s - s_0) = h_2 \lambda \tag{9.17}$$

These are the Laue conditions for interference at two-dimensional lattices as already been used for the geometric theory.

Another consequence of equ. (9.30) is that the heights of the intensity maxima are proportional to $(M_1 \cdot M_2)^2$.

The above derivation justifies the applicability of the geometric theory. Maxima of J are determined by the lattice vectors a_1, a_2 of the unit cell and give, with the aid of the Laue equations, the directions of the interference maxima, i.e. the positions of spots in the diffraction patterns. The intensities of these spots then follow from the magnitude of the other term in equ. (9.29), namely the square of the structure factor F.

Equ. (9.30) can further be used to obtain information about the angular spread $\delta\varphi$ of the diffracted beams and thereby on the diameter of spots (of course without the inclusion of any instrumental effects on the spot diameter, as discussed in section 9.7.2). For a one-dimensional case the interference function is shown in fig. 9.21. This function and therefore also the intensity becomes zero if

$$\sin\left[\tfrac{1}{2} M_1 \boldsymbol{a}_1 \cdot (\boldsymbol{k} - \boldsymbol{k}_0)\right] = 0 \tag{9.32}$$

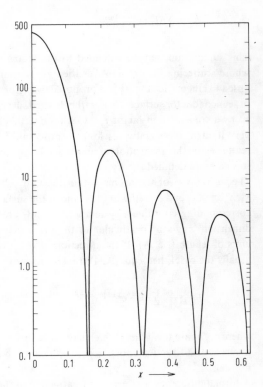

Fig. 9.21. The interference function (9.30) for a one-dimensional lattice consisting of $M_1 = 20$ scatterers as a function of $x = \boldsymbol{a}_1 \cdot (\boldsymbol{k} - \boldsymbol{k}_0)$. Note the logarithmic scale of the ordinate! After Kittel [45].

Hence it can be concluded [48] that

$$\delta\varphi \approx \frac{\tan\varphi}{2 h_1 M_1} = \frac{\lambda}{2 M_1 \boldsymbol{a}_1 \cos\varphi} = \frac{\lambda}{2 d \cos\varphi} \tag{9.33}$$

where $d = M_1 \boldsymbol{a}_1$ is the diameter of the periodic region, which means that the spot diameter (as determined by $\delta\varphi$) is inversely proportional to the number of scatterers within a periodic array. In the ideal case of infinite periodicity the angular spread of the diffracted beams should be negligibly small. In fact this is not observed because of the finite coherence width of the electron radiation available experimentally. This problem is treated in section 9.7 together with the information which may be obtained from the lateral extensions of diffraction spots.

9.6.3. Kinematical structure factor

In the kinematic approximation only those atoms which are (either completely or partially) directly exposed to the incident electron beam are responsible for the scattering intensity. The structure factor then results from a summation over all waves caused by scattering of the primary incident electron wave at the individual atoms ($j = 1 \ldots s$) within the unit cell. If the positions of these atoms are given by r_j and their (atomic) scattering factors by f_j then

$$F = \sum_{j=1}^{s} f_j \cdot e^{i(k - k_0) \cdot r_j} \tag{9.34}$$

This summation may be extended to the atoms within the second and third atomic layers whose scattering factors however then are partially attenuated by the atoms above. Such a "pseudo kinematic" theory was proposed by Lander [17a] and was used to analyse the structures of semiconductor surfaces (with rather limited success). The effect of "shadowing" is incorporated by a transmission factor $t_j \leqslant 1$. Although this concept seems to be plausible it is only of very limited applicability as long as no detailed information about the degree of forward scattering or the absorption of an atomic layer are available. Such data can only be obtained from a more detailed theory.

The positions r_j of the atoms within the unit cell are usually expressed by the coordinates x_j, y_j, z_j, (x_j and y_j being in the plane of the surface). The lattice vectors a_1, a_2 determine the corresponding coordinate system whose units are given by the lengths a_1 and a_2. The third dimension, z_j, is perpendicular to the plane determined by a_1, a_2 and is given in absolute units of length (i.e. Å). If the direction of the primary beam is normal to the surface (as is usually the case) then equ. (9.34) transforms into

$$F_{h_1 h_2} = \sum_{j=1}^{s} f_j \cdot e^{2\pi i [h_1 x_j + h_2 y_j + (1 + \cos \varphi) z_j / \lambda]} \tag{9.35}$$

where h_1, h_2 are the indices which have been introduced already in section 9.5.3 to characterize the various diffraction spots, and φ is the angle between the diffracted beam and the surface normal.

A formal difference between this treatment and the geometric theory must be borne in mind. If a surface structure b_1, b_2 exists, in the geometric theory the diffraction spots are indexed with respect to the reciprocal lattice of the *substrate*, thus leading to fractional order spots. If intensities are calculated with the aid of equ. (9.35) the spot indices are referred to the unit cell of the *surface structure*, as well as the coordinates of the atoms within the unit cell.

If the kinematic theory were valid then the structure factor and consequently the spot intensities could be evaluated for an assumed model of the surface structure using equ. (9.35) which could then be compared with the experimentally determined values. Although such attempts were made in the initial period of LEED such a procedure is not justified. However it is possible to use equ. (9.35) to evaluate characteristic extinctions of certain diffraction spots which may occur for an assumed structure model. Diffraction spots which are kinematically "forbidden" should not have appreciable intensities due to dynamic effects [49].

A yet unresolved question concerns the atomic scattering factor f_j. To a crude approximation it can be described by a relation derived for a spherical charge distribution [86]

$$f(\lambda, \varphi) = \frac{me^2}{2h^2} (Z - f_x) \cdot \frac{\lambda^2}{\sin^2 \varphi} \qquad (9.36)$$

Z is the nuclear charge and f_x the atomic scattering factor for x-rays.

Since $f_x \ll Z$ the atomic scattering factor for electrons should be much larger than that for x-rays, which obviously is true. Furthermore f should increase with λ^2 or, in other words, the scattered intensity should decrease with increasing electron energy, which again agrees with general observations.

In fact however the scattering properties of atoms in a crystal lattice cannot be approximated by such a simple picture but have to be calculated on the basis of the local electrostatic potentials. The atomic scattering factor is then usually described by a series expansion of the type

$$f(\lambda, \varphi_g) = \sum_{l=0}^{\infty} (2l+1) \, P_l (\cos \varphi_g) \, e^{i\delta_l} \sin \delta_l \qquad (9.37)$$

φ_g is the angle between the incident and the diffracted beam and P_l are the Legendre polynomials, as used for example also in the description of the angular part of the electron wave functions in a hydrogen atom. $l = 0, 1, 2 \dots$ denotes "s-", "p-" and "d-wave" scattering, respectively. The scattering properties are then contained in the energy dependent phase shifts δ_l. This formalism is also used in the dynamic scattering theory.

9.6.4. Intensity-voltage (I/U) curves

So far only scattering at two-dimensional objects has been discussed which yields the positions of diffraction spots as the main result. However the intensities of the diffracted beams depend strongly on the periodicity of the crystal normal to the surface, since the primary electrons have finite depth of penetration into matter. These effects manifest themselves in a variation of the structure factor $F(\mathbf{k}, \mathbf{k}_0)$ with the electron energy.

In the three-dimensional kinematic theory (as is applied to x-ray diffraction) the periodicity in the z-direction requires the fulfilment of the third Laue condition

$$\tfrac{1}{2} \mathbf{a}_3 \cdot (\mathbf{k} - \mathbf{k}_0) = h_3 \qquad (9.38)$$

as well as equ. (9.17) in order to obtain intensity maxima. That means that the intensities in the directions as determined by equ. (9.17) exhibit only noticeable amounts if equ. (9.38) is also fulfilled. If c_z is the length of periodicity normal to the surface and the primary beam strikes the surface under normal incidence then for example the 00-beam should have maximum intensities at electron energies U_B which are given by

$$2c_z = h_3 \lambda = h_3 \cdot \sqrt{\frac{150.4}{U_B}} \qquad (9.39)$$

Intensity maxima as determined by this equation are called 'primary Bragg-peaks'.

Their appearance can also be understood on the basis of the Ewald construction. For a three-dimensional lattice the reciprocal lattice 'rods' as introduced for the two-dimensional case (fig. 9.13) are replaced by reciprocal lattice points. Interference phenomena occur only when the Ewald sphere intersects one of these reciprocal lattice points, i.e. at definite wavelengths. If scattering takes place at a two-dimensional lattice no variation of the spot intensities with the wavelength should take place, provided of course that the atomic scattering factors f_j are independent of energy.

The actual situation with LEED is somewhere between these two extreme cases. The sharply defined energies for which interference occurs as determined by the third Laue equation are replaced by energy regions. These regions become broader as the depth of penetration of the electrons decreases since this reduces their "feeling" of the periodicity in the third dimension. An experimentally derived I/U-curve [50] is reproduced in fig. 9.22. The positions on the energy

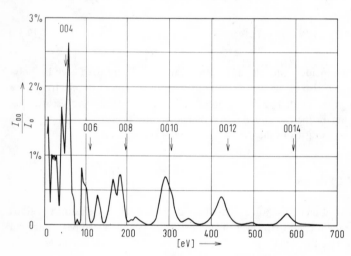

Fig. 9.22. Intensity/voltage (I/U)-curve for the $(0,0)$-beam of Ni(100). The diffracted intensity I_{00} is divided by the intensity of the primary beam I_0. After [50].

scale at which the primary Bragg peaks are expected are indicated by arrows. The actual intensity maxima are shifted by several eV towards lower energies from these positions. The reason for this is that the wavelength of the electrons is altered within the crystal by an "inner potential" U_0. An electron which enters a solid gains additional kinetic energy from the inner potential and has therefore effectively a shorter wavelength

$$\lambda_{\text{crystal}} = \sqrt{\frac{150}{U - U_0}} \ [\text{Å}] \tag{9.40}$$

as compared with the wavelength in vacuum $\lambda_{\text{vacuum}} = \sqrt{\dfrac{150}{U}}$. U_0 lies between -10 and -30 eV and depends on the energy of the incoming electron. U_0 can be derived from the shifts of the Bragg peaks in the I/U-curves. The existence of an inner potential is also responsible for the refractive index $n = \dfrac{\lambda_{\text{vacuum}}}{\lambda_{\text{crystal}}} = \sqrt{\dfrac{U - U_0}{U}}$ which is not equal to unity and is therefore the physical reason why slow electrons are partially reflected elastically at solid surfaces and do not penetrate completely into the solid.

As can be seen from fig. 9.22 not only primary Bragg peaks are present in the I/U-curves. The additional peaks are called secondary Bragg maxima and are caused by multiple scattering effects. They provide an ample demonstration of the nonvalidity of the kinematic theory.

9.7. Disordered structures

9.7.1. Introduction

With the exception of the domain structures as described in section 9.5.5 all the cases discussed so far have been restricted to perfect two-dimensional periodic arrangements of the scatterers on the surface. Such perfection is never found in a real crystal but is subject to lattice imperfections and thermal vibrations as the main disturbing influences. The latter effect leads as in the case of x-ray scattering at crystals to a continuous weakening of the diffraction spots with increasing temperature, combined with an increase of the background intensity. A detailed discussion is given in chapter 10. Deviations from the two-dimensional lattice periodicity may cause characteristic variations of the LEED pattern, although on the other hand it is well known that this method is not very sensitive towards such effects. For example Jona [51] found that about 20% of the atoms on a Si(111) surface may be in a random distribution without causing any detectable change in the LEED pattern.

As outlined in section 9.5.5 the type of variation of a LEED pattern due to partial disorder depends on the mean size d of the regions with uniform periodicity. This section is concerned with those cases where d is smaller than the coherence width of the electrons, L, leading to superpositions of the scattered amplitudes in contrast to the case of domain structures, where $d > L$ and the intensities are superimposed.

9.7.2. The diameter of the coherence zone

Inspection of the interference function J (equ. (9.30)) shows that the width of a beam profile (i.e. the angular spread of a diffracted beam) decreases with increasing numbers M_1, M_2 of scatterers in a periodic array. In the case of a completely perfect surface the diffraction spots should become infinitely sharp. In practice the LEED spots always have a finite size due to the limited diameter of the coherence zone of the electrons, which is caused by the experimental limitations of the system. Under ideal conditions, namely with a perfectly monoenergetic and parallel electron beam, the coherence width would be equal to the diameter of the electron beam (~ 1 mm). In fact such conditions cannot be realized leading to incoherence in time and space which decreases the coherence width [17a], [49], [52]. A detailed discussion of all effects which may contribute to the "LEED response function" has been given by Park et al. [53]. Among the factors to be considered are the following: energy spread (leading to incoherence in time), lateral extension of the electron source, width of the aperture, deviations from parallelity. The latter factors cause incoherence in space. The main effects are caused by the diameter of the electron source leading to a divergence of the electron beam by an angle β_s, and by the thermal spread ΔU of the energy. The diameter of the coherence zone L is then approximately given by [54]

$$L \approx \frac{\lambda}{2\beta_s (1 + \Delta U/U)} \tag{9.41}$$

For energies above 50 eV the energy spread may be neglected and therefore in most cases the coherence width is limited by the diameter of the primary beam [53].

Taking typical data (source diameter 0.5 mm at a distance of 10 cm from the surface, $\lambda = 1$ Å) a value for L of about 100 Å results. This is in agreement with the results from experimental investigations with a Pd(110) surface [55]. In general the coherence width may be assumed to be in the range of 50–100 Å. The size of the surface regions for which information on deviations from the two-dimensional periodicity can be obtained from the LEED patterns is thus fixed. If the mean diameter d of these regions is smaller than L, then the shape of the LEED spots is determined by d, and otherwise by the coherence width L. This is also the reason why excellent LEED patterns are frequently observed from surfaces which are heavily distorted on a macroscopic scale.

9.7.3. Size effects

As can be seen from equ. (9.30) the analytical form of the interference function J determines the shape of the diffraction spots via the angular spread $\delta\varphi$ of the intensity maxima. $\delta\varphi$ is inversely proportional to the number of scatterers in a periodical array (equ. (9.33)). This so-called size effect is a well-known phenomenon in x-ray diffraction [48, 52] and is used for example for the determination of the diameter of very small particles causing characteristic broadening of the interference features.

In a regular arrangement of scatterers M_1, M_2 a situation may occur where $M_1 \gg M_2$. According to equ. (9.33) the diffraction spots should be enlarged in one direction of the reciprocal lattice much more than in the other, leading to streaked patterns such as have been observed in LEED studies quite frequently. Adsorption of oxygen on Cu(110) for example produces the formation of such streaks at low coverages as shown in fig. 9.23 [37]. The adsorbed oxygen atoms are

Fig. 9.23. Formation of streaks in the diffraction pattern of Cu(110) in the early stages of oxygen adsorption [37].

obviously very mobile in the [10] – direction (the furrows on the (110) surface) and agglomerate to equally spaced chains whereas only a very weak correlation between the rows of adsorbed atoms exists in the [01]-direction. If the coverage is increased ordering then takes place in this latter direction accompanied by a gradual contraction of the streaks into spots of a 2×1-structure (fig. 9.15 a).

Related effects are observed after the formation of thin graphite overlayers on metal surfaces. As can be seen from fig. 9.24 this may be associated with the appearance of a ring-like diffraction

Fig. 9.24. Ring-like diffraction pattern caused by a graphite overlayer on Ni(110) [56].

pattern. The explanation is that the base planes of the individual graphite crystallites are parallel to the surface but randomly rotated around the c-axis. In the case of graphite on Ni(110) [56] the intensity of the diffraction rings is not uniform which is due to the occurrence of two preferred orientations on the surface.

9.7.4. Substitution disorder

The formation of domain structures in ordered adsorbed phases can be understood from a consideration of the individual steps involved in the formation of an adsorbate layer. Initially particles arrive on the surface at random from the gas phase and then diffuse on the surface finally joining together and forming ordered aggregates (as in three-dimensional nucleation). With increasing total coverage these islands finally coalesce but since their growth started at completely separate nuclei it is quite possible that antiphase domains are formed at the boundaries (fig. 9.25). This is a consequence of the effect of 'registry degeneracy' [57], i.e. the possibility that the surface structures may have their origin at different (but equivalent) points on the substrate which cannot be connected with each other by multiples of the basis vectors b_1, b_2 of the surface structure. Three cases of this type which can occur with a c2 × 2 structure are illustrated in fig. 9.25. In fig. 9.25a the two domains are connected by a vector $l = b_1 + 2b_2$. Consequently both domains scatter in phase and cause sharpening of the LEED spots due to the enlarged effective diameter of the periodic area as compared with that of the individual domains. In fig. 9.25b and c the relation $l = n_1 b_1 + n_2 b_2$ (n_1, n_2 = integers) no longer holds and in both cases the domains are out of phase. The resulting effect on the diffraction pattern

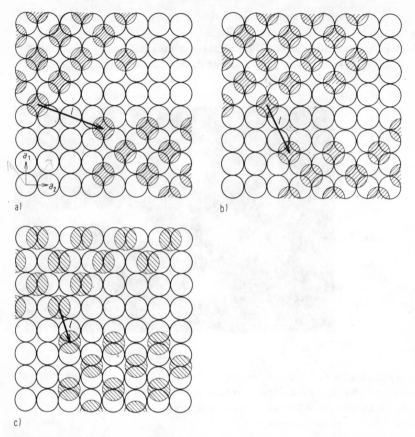

a)

b)

c)

Fig. 9.25. Different types of antiphase domains (substitution disorder). In contrast to fig. 9.17 the domain diameters are now assumed to be smaller than the coherence width. The unit vectors of the surface structure are given by $b_1 = a_1 + a_2$, $b_2 = a_1 - a_2$. a) The two domains are connected by $l = -b_1 + 2b_2$. b) $l = a_1 - 2a_1$. Only the 'extra' spots will be split or enlarged. c) $l = \frac{1}{2}a_1 - \frac{3}{2}a_2$. All diffraction spots will be affected.

may best be illustrated by assuming two domains of equal size [17e]. The interference function is then given by

$$J' = 2J\{1 + \cos[(k - k_0) \cdot l]\} \tag{9.42}$$

where J is the normal interference function (9.30) for an arrangement with uniform periodicity. Maxima in J occur if $(k - k_0)/2 = h_1 b_1^* + h_2 b_2^*$.

It is then possible for the term in brackets to disappear. This happens if

$$h_1 l_1 + h_2 l_2 = \frac{1}{2}(2n + 1) \qquad (n = \text{integer}) \tag{9.43}$$

where l is written as $l = l_1 a_1 + l_2 a_2$.

The consequence of this can be demonstrated with the c2 × 2-structure as drawn in fig. 9.25b, where $l_1 = 1$ and $l_2 = -2$. Possible diffraction spots are indexed by h_1, $h_2 = \frac{1}{2}m_1$, $\frac{1}{2}m_2$. If m_1,

m_2 are both odd this leads to 'extra' spots; if both are even 'normal' (substrate) spots are the consequence. In this case equ. (9.43) is fulfilled for spots for which m_1 and m_2 are odd, and all 'extra' spots are split. If the domains are not of equal size perfect splitting does not occur but rather a broadening of the 'extra' spots. The substrate spots are not affected as long as l is a vector of the substrate lattice, which is always the case if the coordination of the adsorption site has the same degree of symmetry as that of the atoms of the substrate surface. In the above example this condition is fulfilled if the adsorbed particles are either placed on top of a substrate atom or equally spaced between four substrate atoms. On the other hand, in fig. 9.25c adsorption has taken on sites with twofold symmetry, and l is not necessarily a vector of the substrate lattice so that the existence of antiphase domains now leads also to splitting of the substrate diffraction spots.

These effects provide a unique method of determining the coordination of adsorbed particles with respect to the surface atoms, which would otherwise only be possible after a complete analysis of the intensities.

A quantitative treatment of the effects of spot splitting must be tackled on the basis of the statistics of occupancy of the existing adsorption sites. The resulting LEED pattern may then be evaluated by means of the kinematic theory.

A detailed theory of this type has been derived by Houston and Park [58] for a one-dimensional model. The situation is illustrated by fig. 9.26. In this example it is assumed that within a

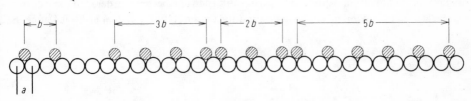

Fig. 9.26. One-dimensional model for imperfect domain structures. The domain length is defined in terms of the number of "normal" spacings b. After Houston and Park [58].

domain the mutual distance b between two adsorbed particles is given by $b = 2a$, leading to half order spots in the corresponding diffraction pattern. At a given coverage θ/a a statistical distribution of domains with various domain lengths will exist. The result reveals that for $\theta < 1/2$ the half order 'extra' spots should split continuously with decreasing coverage. The distance between the divided spots in the diffraction pattern should vary linearly with coverage. Such phenomena have been observed experimentally with a number of systems [59]–[61], [34]. For example the LEED patterns due to oxygen adsorption on a Pd(110) surface [34] are shown in fig. 9.27, where a 1×3-structure is continuously transformed into a 1×2-structure with increasing coverage.

Earlier attempts [59] to explain such effects were based on an assumption of uniform mixing of different structural elements [62] (for the example shown in fig. 9.26 these elements would have periodicities of $2a$ and $3a$). But in fact considerable deviations from a uniform distribution may be tolerated before the splitting of spots is destroyed [58].

The one-dimensional model has also been developed further [63] and leads in general to the same results for spot splitting.

A more realistic two-dimensional model was developed by Ertl and Küppers [64] on the basis of the intermolecular interactions existing between adsorbed particles and using computer

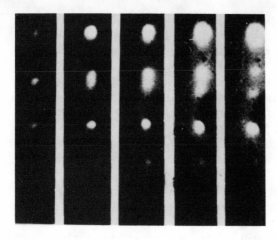

Fig. 9.27. Diffraction patterns for the system O/Pd(110). With increasing coverage the 'extra' spots continuously coalesce [34].

simulation by means of the Monte Carlo method. A completely disordered distribution causes a continuous increase in the background intensity in the LEED pattern as usually observed with disordered adsorbed layers. If then the operation of certain interaction energies between neighbouring adsorbed particles (either attractive or repulsive) is assumed, the equilibrium configuration is attained by surface diffusion and becomes ordered to some degree. Fig. 9.28 shows the equilibrium configuration for the Pd(110)/O_2 system at $\theta = 0.4$ after the

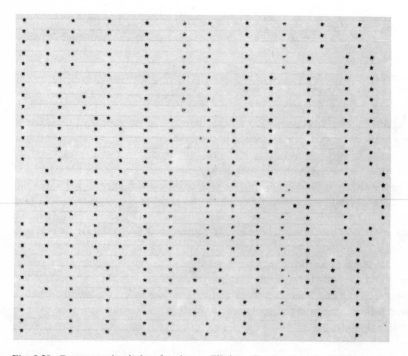

Fig. 9.28. Computer simulation for the equilibrium distribution of adsorbed oxygen atoms on Pd(110) at $\phi = 0.4$.

diffusion process with suitably chosen interaction parameters (which were qualitatively justified by measurements of the isosteric heats of adsorption).

Although this configuration is still considerably disordered, the calculation of the corresponding LEED pattern with the computer technique as described in section 9.9 revealed the appearance of 'extra' spots at positions h_1, $(h_2 \pm 0.4)$ of the diffraction pattern. The result of evaluations for different coverages is reproduced in fig. 9.29 and shows the splitting of spots which varies linearly with the coverage, as observed experimentally.

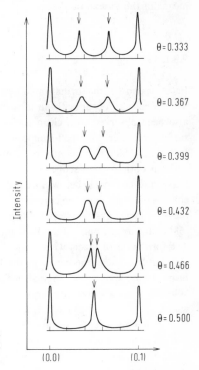

Fig. 9.29. Calculated continuous beam splitting for the system O/Pd(110). Relative LEED intensities between the (0, 0) and (0, 1) substrate spots in [01]-direction of the reciprocal lattice.

Fig. 9.28 demonstrates the insensitivity of the LEED method towards deviations from the ideal two-dimensional periodicity. This fact together with the limited coherence width of the electrons is the reason why LEED patterns may be obtained from many 'imperfect' surface structures.

It must also be pointed out that spot splitting and continuous shifting of 'extra' spots may also be caused by a completely different mechanism, namely if the unit cell of an adsorbate structure is continuously compressed upon further increasing the coverage. This interpretation has been used for several experimentally studied systems [65].

To decide which of the two reasons for the observed spot splitting (substitution disorder or compression of the unit cell) is operating in a particular case presents usually no fundamental problem, especially if further information is available, for example the size of the adsorbed particles or the variation of the adsorption energy with coverage, etc. Compression of the unit cell is usually associated with the appearance of additional spots due to multiple scattering, since these compressed structures are necessarily incoherent.

9.8. Facets and stepped surfaces

The effects which were treated in the preceding section were due to partial disorder of the surface structure on a perfect substrate lattice, but it is also possible that the substrate structure itself is modified and is no longer strictly periodic. This does not mean that in some cases the unit cells of the topmost layers of a substrate lattice differ from the bulk structure, but refers to the occurrence of faceting and step arrays on a surface. These effects may also be treated by means of the kinematic theory.

9.8.1. Faceting

Faceting denotes the formation of new crystal planes which are inclined to the original surface. This effect can occur with an adsorbate covered surface at elevated temperatures, but has also been observed with clean surfaces [66] owing to the tendency of a crystal to lower its surface free energy by the formation of new planes with different crystallographic orientations [67]. These facets frequently have macroscopic dimensions and can therefore be identified with an optical microscope. It is vital to realise that the formation of facets may often be induced at coverages below a monolayer (in particular by oxygen), which indicates that this effect is different from the corrosive formation of three-dimensional compounds. Examples of the latter type have also been studied by LEED, for example the formation of NiO on Ni(111) [68] or the appearance of $\alpha - W_2C$ pits on a W(111) surface [69].

The formation of facets with surfaces which are inclined to the nominal surface orientation may be detected immediately in the LEED pattern. With normal incidence of the primary beam the 0,0-spot of the original surface will be located in the center of the diffraction screen. With increasing electron energy all other spots move continuously towards the 0,0-spot, whose position remains fixed. If a plane is inclined by an angle $\neq 90°$ to the incident beam the corresponding specular beam (i.e. 0,0-spot) will be at a different position of the diffraction pattern or even out of the picture. With increasing electron energy the diffraction spots from the facet surface will move continuously towards the corresponding specular beam and therefore not to the center of the diffraction pattern.

The Ewald construction of fig. 9.30 reveals that the reciprocal lattice rods of the normal and of the facetted surface intersect each other in a series of points. If with variation of the electron energy the Ewald sphere passes through these points the diffraction spots of both surface orientations will coincide in the diffraction pattern. Tucker [70] has evaluated a simple treatment in order to identify the orientations of facets using this effect.

If a beam originating from a facet passes the normal 0,0-beam the following relation holds (fig. 9.30):

$$n\lambda = 2d_F \sin \alpha \tag{9.44}$$

where α is the angle of inclination of the facet, d_F the periodicity within the facet surface and n their order of diffraction. The condition for interference for the faceted surface is

$$d_F (\sin \varphi - \sin \alpha) = n\lambda \tag{9.45}$$

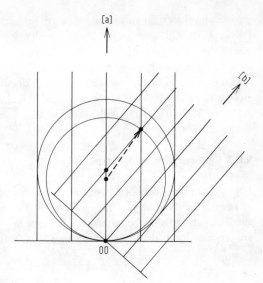

Fig. 9.30. Ewald construction for the reciprocal lattices of the normal surface (direction [a]) and of a facet (direction [b]). The dashed line indicates the situation where the Ewald sphere passes through a point common to both reciprocal lattice rods. These points of intersection are reciprocal lattice points.

from which follows

$$\frac{\mathrm{d}\lambda}{\mathrm{d}\varphi} = \frac{d_\mathrm{F}}{n} \cdot \cos \varphi \tag{9.46}$$

If this beam passes through the normal 0,0-spot:

$$\frac{\mathrm{d}\lambda}{\mathrm{d}\varphi} = \frac{d_\mathrm{F}}{n} \cdot \cos \alpha \tag{9.47}$$

This equation combined with equ. (9.44) yields

$$\frac{\mathrm{d}\lambda}{\mathrm{d}\varphi} = \frac{\lambda}{2} \cdot \cot \alpha \tag{9.48}$$

Therefore it is possible to determine the angle of inclination α from the variation of the angle of diffraction φ with the wavelength λ, when passing through the normal $(0,0)_\mathrm{N}$-spot. d_F then can be determined with the aid of equ. (9.44). If the order of diffraction n is unknown, it is also possible to measure the change of the reciprocal wavelength $\Delta(1/\lambda)$ with the passage of various facet beams with different n through $(00)_\mathrm{N}$:

$$\frac{\Delta n}{\Delta(1/\lambda)} = 2 d_\mathrm{F} \sin \alpha \tag{9.49}$$

A complete crystallographic indexing of the faceted plane can be achieved by using the fact that the points of intersection of both sets of reciprocal lattice rods are points of the three-dimensional reciprocal lattice of the crystal [68], [71]. This results in the intensities reaching maximum values if the Ewald sphere passes through these points, since the third Laue condition (the depth condition) is then satisfied for both planes. The construction of the three-dimensional reciprocal lattice can now in turn be used to calculate those electron energies which are associated with the intense primary Bragg maxima [71].

The formation of facets has been studied in detail with a series of systems, particularly with oxygen interaction on various W surfaces [71, 72], on Rh(210) [70] and Cr(110) [73], and also for Al on Si(100) [74] and with clean GaAs(111) surfaces [66].

9.8.2. Surfaces with regular steps arrays

High index planes may be considered as consisting of terraces of low index planes with steps of equal length and height. Such a situation may occur if, for example, a crystal is cut with a small deviation from the correct angle for a low index plane, or if semiconductor surfaces are prepared by cleavage. As is clear from fig. 9.31 these cases represent a combination between

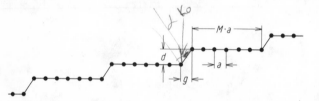

Fig. 9.31. Cross section of a surface consisting of terraces with regular steps array.

faceting (the side planes of the terraces are inclined) and antiphase domains with equal lengths of periodicity so that both facets and antiphase domains have similar diffraction properties. They both give spot splitting at certain electron energies, and different specular spots towards which the patterns converge with increasing electron energy. The first systematic studies with such systems were performed by Ellis and Schwoebel[75] with a UO_2-surface which was inclined by $11.4°$ towards the (111)-plane. A typical LEED pattern with doublet spots is shown in fig. 9.32. Irregular spacing between the steps leads to the formation of streaks instead of split

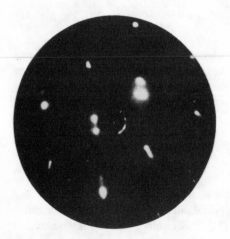

Fig. 9.32. LEED pattern from a UO_2 surface inclined $11.4°$ from the (111) plane showing doublet spots. $U = 75$ eV. After Ellis and Schwoebel [75].

spots similar to antiphase domains with non-equal diameters. Similar observations were reported by Lang et al. [22] for Pt high-index planes (these authors proposed also a nomenclature for ordered stepped surfaces) and by Henzler[76] for cleaved Ge(111) surfaces. Further LEED investigations with stepped surfaces have been made with Nb[77] and Re[78]. Rhead and Perdereau [79] performed detailed studies with so-called "vicinal planes" of Cu, where besides regular step arrays additional effects due to regular spaced kinks were observed. Following Henzler [76] LEED patterns from stepped surfaces may be analyzed as follows:

Assuming an infinite step array (i.e. at least 20 equally spaced steps) of the kind as drawn in fig. 9.31 the interference function for a one-dimensional model becomes

$$I(\varphi) = \frac{\sin^2\left[\frac{1}{2}ka(M+1)\sin\varphi\right]}{\sin^2\left[\frac{1}{2}ka\sin\varphi\right]} \cdot \sum_{m=-\infty}^{+\infty} \delta\left[\frac{1}{2}k(Ma+g)\sin\varphi + \right. $$
$$\left. + \frac{1}{2}kd(1+\cos\varphi) - m\pi\right] \tag{9.50}$$

$M+1$ is the number of atomic rows on a terrace plane; the distances a, d and g are evident from fig. 9.31; $k=2\pi/\lambda$. The first term is the normal interference function of a periodic array consisting of $M+1$ atoms and exhibits maximum values if $\frac{1}{2}ka \cdot \sin\varphi = n\pi$, i.e. these are the normal substrate spots. The second term represents a sum of δ-functions (the summation extends to infinity since an infinite array was assumed), which are non-zero in those cases where their arguments disappear. This term depends only on the width $(Ma+g)$ of the terraces and on the step height d. Using equ. (9.50) the separation between two δ-functions and thereby the separation between two (split) spots is

$$\Delta\varphi = \frac{\lambda}{(Ma+g)\cos\varphi - d\sin\varphi} \tag{9.51}$$

which simplifies in the vicinity of the (0,0)-spot ($\varphi=0$) to

$$\Delta\varphi_{0,0} \approx \frac{\lambda}{Ma+g} \tag{9.52}$$

The terrace width $(Ma+g)$ can then be determined from the magnitude $\Delta\varphi_{0,0}$ of the splitting of the 0,0-spot. A δ-function will always be found at a position of the diffraction pattern corresponding to specular reflection of the inclined surface, given by

$$\varphi_s = \frac{2}{\tan\left(\dfrac{d}{Ma+g}\right)} \tag{9.53}$$

This position can be determined by variation of the electron energy and yields the step height d. Upon variation of the electron energy the δ-functions also tend to pass through the positions of the normal spots, so that at certain wavelengths these are not split. This situation occurs for the 0,0-beam if

$$U_{0,0} = \frac{150\, s^2}{4\, d^2} \tag{9.54}$$

where s is an integer and d the step height. For half-integral values of s the 0,0-beam will be split symmetrically into two spots with equal intensities. Using this relation Henzler [76] found that on cleaved Ge(111) surfaces only steps with the minimum height allowed by the crystal structure $d=3.27$ Å appear. The terrace width could vary between 5 and 15 atomic rows on different parts of the sample surface, which leads to different splitting of the spots. The analysis so far has been based on the assumption of the existence of perfectly ordered steps. In fact a more or less statistical distribution of steps will be present. Houston and Park [80] developed a kinematic formalism which includes the occurrence of statistic step distributions. The most striking result is that a relatively high degree of scatter in the step lengths still produces sharp splittings in the diffraction pattern, which is in agreement with the general result that LEED is rather insensitive to structural imperfections.

More recently Laramore et al. [81] investigated the influence of a relatively small number of steps on the LEED intensity profiles (I/U-curves). It is shown that a step distribution reduces the central beam intensities (with respect to the perfect surface case) and also shifts the Bragg peaks in the I/U-curves to somewhat higher energies. This could become of special relevance for adsorbate covered surfaces with antiphase domains, which are a special case of this effect.

9.9. Simulation of diffraction patterns

In the more complex cases, such as with disordered structures, antiphase domains and step arrays as discussed in the preceding sections, it is often useful to assume a certain model for the surface and then to evaluate the corresponding diffraction pattern by means of the kinematic theory which is then compared with the experimentally observed LEED pattern.

If a set of scattering particles is distributed (more or less at random) over a periodic array of substrate sites the amplitude of the scattered wave again is given by equ. (9.26). However the important fact is that the atomic scattering factors f_j now may have two different values (f_j or f_0), depending on whether the site j is occupied by an adsorbed particle or not. The intensity then can be calculated by means of a computer program as developed by Ertl and Küppers [46] for a substrate lattice with 30×30 sites. This method can also be extended to other models (e. g. step arrays). The choice of a lattice consisting of 30×30 sites is justified by the coherence width of the electrons which is of this order of magnitude, so that superposition of scattered amplitudes only takes place within areas of this size. An application of this technique is discussed in section 9.7 (spot splitting with partially disordered structures).

Another application of this technique is to calculate the relative spot intensities from partially disordered structures. It is also possible to follow order-disorder transitions in the adsorbed phase. Since the relative intensities at fixed electron parameters (k, k_0) are considered, the only essential deviation from the results of a complete dynamic LEED theory is that within this model the structure factors of the individual surface unit cells do not depend on the environment (i. e. whether the neighbouring sites are occupied or not). The first attempts to include the effects of disordering into a dynamic scattering theory were made recently by Duke and Liebsch [82].

If the diffraction pattern only (=interference function) is desired a straightforward technique may be used which is based on the evaluation of the optical transform of an assumed structure model. Such methods have been proposed by Ellis and Campbell [83] and by Fedak et al. [84]. At first a two-dimensional diffraction grating is prepared by plotting the structure model consisting of about 30×40 scatterers on a sheet of paper (format about 20×30 cm^2). Fine holes are then pricked through with a needle at the positions of the scatterers. Various atomic scattering factors may be simulated by holes of different diameters. The paper is then put in front of an illuminated translucent screen and photographed to reduce the image size to about 4×6 mm^2. The dimensions of the scatterers are then in the range of the wavelength of visible light so that an optical diffraction pattern may be produced with coherent light from a laser.

As an example the splitting of "extra" spots of a c2×2-structure by parallel antiphase domains is illustrated by fig. 9.33 [85]. The patterns shown in fig. 9.33a)–d) are sharp since the gratings are large and the surfaces are fully covered. At lower coverages there may be nucleation arrays consisting of only two antiphase domains (fig. 9.33e) leading to more diffuse diffraction spots. On an actual surface two types of domain orientations rotated by 90° to each other are to be expected and would give rise to a LEED pattern as reproduced in fig. 9.33f.

9.10. Elements of a dynamic LEED theory

9.10.1. The need for a dynamic theory

In the preceding sections reference was mainly restricted to the interference function using the kinematic theory to derive information about spot positions and shapes, and these results are valid even in cases where the kinematic theory is not applicable. However a complete structural analysis including for example the positions of adsorbed particles with respect to the substrate atoms is only possible via the structure factor F.

The kinematic theory is based on the assumption that all atoms scatter independently and are excited only by the primary (i.e. incoming) wave. Owing to the large values for the scattering factors predicted by equ. (9.36) it is evident however that this approximation of (weak) single scattering events cannot be valid in the case of slow electrons. Diffraction patterns of the type as discussed in section 9.5.6 show clearly that multiple scattering at two layers with different periodicity has to be assumed in order to explain the observed features. Further experimental evidence for the operation of dynamic effects can be drawn from the I/U-curves as shown in fig. 9.22, where besides the primary Bragg peaks (as predicted by the kinematic theory) further structure and in particular secondary Bragg maxima can be detected. The only exception so far found is a single-crystal Xe-surface, where owing to the particular properties of xenon atoms a purely kinematic intensity profile has been observed [87].

The formation of secondary Bragg peaks can be readily understood on the basis of multiple scattering. If a primary beam propagating in direction s_0 is scattered within a three-dimensional periodic crystal, then interference will take place in the directions s', which obey the condition

$$s_0 - s' = \lambda \tilde{g} \tag{9.55}$$

In this case $\tilde{g}$ represents a reciprocal lattice point (instead of a reciprocal lattice rod), since the third Laue condition must also be satisfied. The scattered beam s' is now able to interfere for a second time in the directions s before it leaves the crystal, if

$$s' - s = \lambda \tilde{g} \tag{9.56}$$

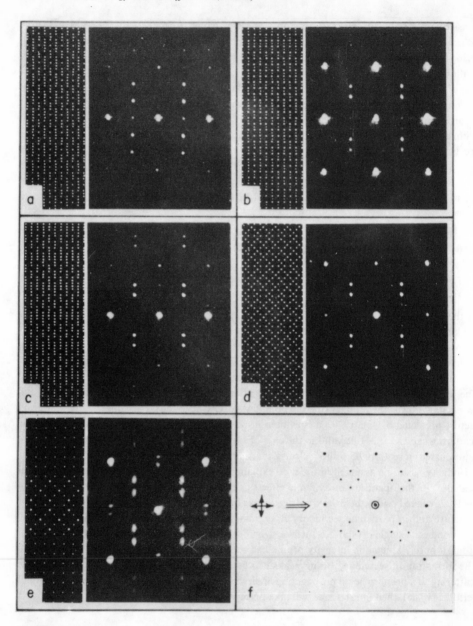

Fig. 9.33. Optical simulation of LEED patterns from parallel antiphase domains of a c 2 × 2-structure. a) Bridge bonds, domains three substrate lattice spacings wide.   b) Same as a), five spacings wide. c) Bridge bonds, zig-zag boundaries, domains five spacings wide.   d) Centred bonds, five spacings wide.   e) Centred bonds, only two antiphase domains.   f) Pattern for two equal probabilities of two domain orientations. After Ellis [85].

The directions of the backscattered beams which arrive at the fluorescent screen are then given by *s*. Fig. 9.34 illustrates such a possibility in the case of the 0,0-beam.

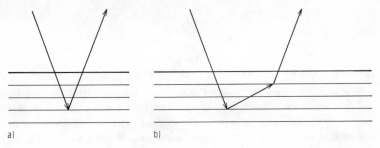

Fig. 9.34. Diagram illustrating the formation of the specular beam a) by single diffraction, b) by multiple diffraction.

For the 0,0-beams of cubic crystals this new maxima in the I/U-curve can be classified as Bragg peaks of fractional order [88]. It has been found that the intense peaks are those which are indexed by multiples of the ratio d/c_z, where d is the distance between the layers and c_z the periodicity in the direction normal to the surface (e.g. for the (111) plane of a fcc crystal is $d/c_z = 1/3$) [89]. Secondary Bragg peaks in general are very sensitive to the presence of surface impurities and imperfections.

A final result from equ. (9.36) for the kinematic theory is that the scattering factor should increase with the atomic number. In this case it would be very difficult to understand why light adsorbed atoms like H or O on substrates consisting of heavy metal atoms (Ni, W etc.) could cause the formation of 'extra' spots in the LEED patterns with intensities which are comparable to those from the substrate, which in fact has been observed for a number of systems. Attempts to explain this difficulty led to the concept of "reconstruction" in which adsorption causes a place exchange of substrate atoms which then give rise to the appearance of intense additional diffraction spots because of their altered periodicity [90]. Although this effect may not be dismissed a priori (for example, faceting is a macroscopic variation of the position of substrate atoms under the influence of adsorbed layers), the 'reconstruction' theory did not have to wait long for rebuttal [91]. Multiple scattering calculations have demonstrated that even a weak scatterer like adsorbed hydrogen may give rise to respectable LEED intensities so that reconstruction of the substrate atoms is not required [92], but no definitive conclusions have yet been drawn and this topic is still the subject of discussion.

Several comprehensive reviews on LEED theory have been published recently [17j, k, l–q]. More than an outline of the ideas behind the various approximations is beyond the scope of the present book, particularly in view of the mathematical formalism sometimes involved.

9.10.2. Physical parameters entering a complete theory

The aim of a complete theory is to calculate the intensity of electrons backscattered from a periodic crystal as a function of primary energy and direction. Usually one is interested in I/U-curves, i.e. the intensity of an individual diffraction spot as a function of energy, which has to be compared with experimentally derived spectra. Starting points are the rules of conservation for energy and momentum:

$$E(k_0) = E(k)$$

and   (9.57)

$$k_{||} = k_{||}^0 + 2\pi g$$

where $k_{||}$ and $k_{||}^0$ are the components of the momenta of the scattered and primary beams parallel to the surface. The latter equation is identical with the interference conditions for diffraction at a two-dimensional lattice. The result can be expected to depend on the following factors: i) the interactions between the electrons and the ion-cores and valence electrons near the surface, ii) the positions and vibrational states of the ion-cores, the latter being of course a function of temperature.

The basis of all theories is the so-called 'static potential model' [93] in which it is assumed that the atoms are rigidly located in a lattice (as used for all band structure calculations). It is however also possible to include lattice vibrations a priori by assuming that the positions of the nuclei are functions of time [94].

Before considering the mathematical formalism which has to be developed in order to solve the multiple-scattering problem the various physical parameters which as well as surface geometry are involved in the process must be recognised.

The Hamiltonian describing the problem may be formulated as follows:

$$H_0 = -\frac{\hbar^2 \Delta}{2m} + V_p + V_0$$ (9.58)

V_0 is a constant potential and $V_p = \sum_n V_n(r - R_n)$ where V_n is the static potential of an atom n, whose nucleus is positioned at R_n.

a) *The scattering potential.* The first difficulty arises from the term $V_n(r - R_n)$ since it represents the self-consistent ion-core potential of a nucleus in a sea of mobile electrons, which of course is different for a solid than for an isolated atom and depends not only on r but also on the energy. The usual approach used in LEED theory is the "muffin-tin" potential [88]. Frequently empirical values are used [92], although some papers also start with a pseudopotential approach [95]. The ion-core electron distribution may best be obtained by a Hartree-Fock calculation; many electron contributions may be included by the use of exchange and correlation terms [96], [97]. In some work the potentials as developed for band-structure calculations were used [98]. As described by equ. (9.37) the atomic scattering factors are expanded into partial waves characterized by phase shifts δ_l. These phase shifts may be derived from the scattering potential.

b) *The inner potential.* The term V_0 in equ. (9.58) describes the fact that the potential energy for electrons inside a crystal is lower than in vacuum – an effect which leads to a variation of the electron wavelength and to energy shifts of the primary Bragg maxima in I/U-curves, as outlined in 9.6.4. Attempts at a theoretical determination of this inner potential have so far been fairly unsuccessful, so that a value is usually obtained from experimental I/U-curves by shifting experimental and theoretical curves on the energy scale with respect to one another until the maxima occur at the same energy.

c) *The variation of the potential at the surface.* Although several derivations of the potential in the bulk of a periodic solid exist and at large distances from the surface the 'image-force' potential provides a useful approximation, only very little information on the shape of the potential just in the region of the surface is available [99]. Opinions differ as to the influence of the potential shape at the surface on the calculated LEED spectra [94], [100], but it turns out that an acceptable approximation for evaluating I/U-curves is to ignore the potential step at the surface altogether [101]. A further point is that the scattering potentials of the topmost layer may differ somewhat from those in the bulk, but experience indicates that these effects are of minor importance.

d) Lattice vibrations. Thermal vibrations of the atoms reduce the intensities of the diffracted beams and increase the background intensity. By analogy with x-ray diffraction effective Debye-Waller factors can be determined experimentally from measurements of spot intensities as a function of temperature as outlined in chapter 10. By use of these factors the intensities calculated for a rigid lattice are then transformed into data for finite temperatures (usually room temperature, where the experimental I/U-curves were taken). Another possibility would be the a priori inclusion of vibrations into the theory [94] and the use of temperature – dependent phase shifts for the description of the scattering potential [101].

e) Inelastic effects. The variations of the LEED intensities with temperature are caused by interactions with lattice vibrations (phonons) giving rise to very small energy-losses (~ 10 meV) which form the "quasi-elastic" part of the elastic peak. Furthermore elastic electrons propagating in a solid interact strongly with the valence electrons, giving rise to both single and collective excitations as described in some of the preceding chapters. This effect becomes evident from the low values of free mean path as drawn in fig. 1.2 and must be included in any serviceable LEED theory [102]. The attenuation length can be included in the formalism either by using a complex wave vector or by introducing an imaginary part of the mean inner potential [109]. The latter is usually assumed to be a few eV, whereas the attenuation length is about 4–10 Å.

9.10.3. Semi-empirical extensions of the kinematic theory

In the kinematic theory the isolated atoms of the lattice are considered as scatterers, each of which is struck by the primary wave field of the incident electrons. In fact these electrons have only a very limited depth of penetration into the crystal. It would appear therefore to be more reasonable to expose the different atomic layers to a primary wave field with attenuation depending on the distance from the surface. This is equivalent to multiplying the atomic scattering factors by an attenuation factor whose magnitude depends on the position of the atom [17a]. It can easily be seen that such a procedure leads to a broadening of the Bragg-peaks in the I/U-curves, as in fact is observed experimentally.

When describing the influence of the third dimension, i.e. the depth condition, it is obviously sufficient to consider complete atomic layers rather than single atoms. Scattering is then treated layer by layer, and each atomic layer has certain reflection and transmission coefficients attributed to it. The coefficients must be adjusted so that the incident wave decreases to negligible values within a small depth. This concept was originally developed by Darwin [103] for the scattering of x-rays and may be extended to include a second scattering of the incident beam at the atomic layers which predicts the occurrence of secondary Bragg maxima in the I/U-curves [104]. Finally this treatment leads to Darwin's dynamic theory [103], which takes all possibilities of scattering between the various atomic layers into account until the procedure becomes self-consistent.

Similar attempts to extend the kinematic theory have been performed by Gafner [105] and by Gerlach and Rhodin [89b]. These techniques are at best able to interprete general features of measured I/U-curves but cannot be used for the analysis of real surface structures.

9.10.4. Multiple-scattering models

A rigorous treatment of the multiple scattering of electrons is based on the following considerations. The individual atoms in the lattice have a certain scattering potential which interacts with an effective wave field. This effective wave field is composed of the primary wave (which is assumed to be a plane wave) and of the sum over all scattered waves originating from the other atoms which reach an atom j. By using this (at first unknown) effective wave field the scattered amplitude from the atom j is calculated which in turn contributes to the scattering at the other atoms. The mathematical treatment has to be continued until the results become self-consistent. Such a formalism was originally developed by Lax [106] and was used by McRae [88] for the solution of the LEED-problem. In the first papers [88, 89a, 107] the atoms were treated as completely isotropic (s-wave) scatterers. Kambe [108] extended this concept to include contributions from higher-order partial waves, i.e. anisotropic scattering effects. (Calculations for W(100) demonstrated in particular the importance of d-wave shifts [92]). However in these earlier papers the importance of inelastic damping was under-estimated.

The operation of strong inelastic effects was emphasized mainly by Duke and Tucker [102, 110] in their 'inelastic-collision' model. These authors argued that due to the rather strong damping of electrons a relatively satisfactory solution will result even if the scattering at the ion-cores is treated only by a relatively crude approximation, namely by assuming only isotropic (i.e. s-wave) scatterers. Their mathematical formalism for the three-dimensional scattering problem was adopted from a theory of Beeby [111]; a similar formalism was developed by Watts [112]. Some simplification of the mathematics is to be expected if the crystal is thought to be composed of individual atomic layers parallel to the surface. It is then convenient to treat the multiple scattering within the layer first and afterwards between the layers. This idea has been developed further by several investigators [113, 114].

An apparently completely different approach begins with band theory. The occurrence of "forbidden" energy ranges in the electronic energy band structure of a solid can be interpreted as being caused by total reflection of the propagating electron waves. It is then only a short step to relate these band-gaps with the Bragg peaks found in electron scattering. This concept stems from an early fundamental paper by Bethe [115] and was first applied to the connection between realistic band structures and I/U-curves by Boudreaux and Heine [116]. In this approach the scattering potential which is periodic in space is represented by its Fourier transform in momentum space. This means that the wave functions in the crystal are composed of superpositions of possible Bloch functions. This solution of the Schrödinger equation for inside the crystal has to be matched at the surface with the general solution outside which consists of an incident and a back-scattered plane wave. The simplest approximation is the "two-beam" case, where only one Bloch wave within the crystal is combined with an incident and a reflected wave. However for realistic derivations it is necessary to include a large number of Bloch waves with complex wave vectors into the calculations (n-beam case). The resulting band structure then contains besides those bandgaps which are associated with the primary Bragg peaks further features which give rise to additional structure in the I/U-curves. This band structure approach has been developed by various investigators and a series of I/U-curves for clean surfaces which agree reasonably well with the experimental data have been obtained [117], [118].

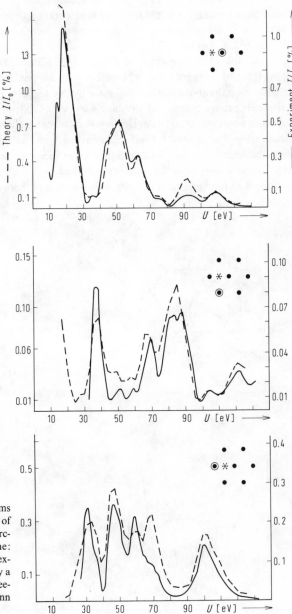

Fig. 9.35. I/U-curves for several beams of Ag(111) at an angle of incidence of 8°. The corresponding spots are encircled. Full line: experiment, dashed line: theory. Note that the scale of the experimental data has to be enlarged by a factor of 1.45 in order to obtain agreement with theory. After Forstmann [17k].

In fact both approaches are equivalent and lead to the same results, as has been shown explicitly by Capart [118] for one particular system.

It seems as if the multiple-scattering method, including the layer by layer treatment, offers a more successful way to treat realistic surface structures. A refined technique has been developed by Pendry [114], [119] in which the multiple scattering within the layers is solved exactly whereas between the layers the back-scattering is treated by perturbation theory. Tait et al. [120] demonstrated that a third-order pertubation calculation yields fairly good results for Al and Ni. At present (i. e. 1973) theory is at a stage where general consensus is attained that the existing models are able to yield quantitative structural data at least for clean metal surfaces [98, 123, 134, 173–176].

As an example of the present state of the art fig. 9.35 shows the results of recent calculations for Ag(111) by Forstmann [17k] together with experimental I/U-curves. These calculations were based on the above method suggested by Pendry [114, 119]. Strong damping was assumed in the energy range considered, the imaginary part of the potential being -4 eV. The scattering potentials were determined from Hartree-Fock wave functions with the inclusion of an exchange contribution. The derived theoretical intensities were multiplied by a Debye-Waller-factor for thermal lattice vibrations as determined experimentally by Lagally [121]. By adjusting the zeros of the energy scale an inner potential of -10.5 eV was determined. As can be seen from fig. 9.35 the agreement between experiment and theory is quite satisfactory.

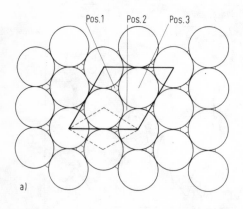

a)

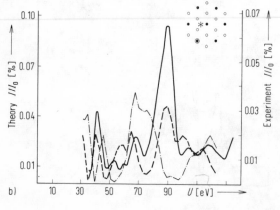

b)

Fig. 9.36. Structural analysis for iodine adsorbed on Ag(111).

a) The unit cell of the $\sqrt{3} \times \sqrt{3} /30°$ overlayer structure and the three possible positions for location of the I atoms.

b) I/U-curves for the (01) beam of iodine covered Ag(111). Full line: experimental. Dashed line: theory for I in position 1. Dot-dashed line: theory for I in position 2. After Forstmann et al. [97].

The first complete structure analyses of adsorbed layers were also made by this technique, namely for sodium on Ni(111) and Ni(100) [100] and for iodine on Ag(111) [97]. The results of the latter investigation are illustrated by fig. 9.36. Adsorbed iodine on Ag(111) forms a $\sqrt{3} \times \sqrt{3}/R\ 30°$ structure with the unit cell as drawn in fig. 9.36a. Three positions for the iodine atoms are possible. By a trial- and error-method best agreement between theoretical and experimental I/U-curves (fig. 9.36b) was found for the adsorbed atoms in position 1 having a distance of 2.24 Å from the silver surface. This leads to an interatomic distance I-Ag (with an accuracy of about 5%) of 2.8 Å which is the same as in crystalline silver iodide. A great deal of work has also been done with simple $c2 \times 2$ overlayers of O [177–179], S [177, 179, 180], Se [177] and Te [177] on Ni(100). In the case of adsorbed oxygen complete agreement between the different groups has not yet been obtained, but nevertheless the results offer optimistic prospects that structure analysis from LEED intensities may become routine work at least for simple adsorbate structures in the near future.

9.11. Averaging methods

At present there are two trends in the attempts to derive surface structures from LEED intensities. The first consists of refining the calculations further in order to obtain better agreement between experimental and theoretical I/U-curves using microscopic dynamic theories. Apart from the few examples mentioned in the preceding section such calculations have been performed only for clean metal surfaces. A large amount of computation is required and at present it is difficult to predict whether this method will ever be generally applicable to routine analysis of surface structures.

Another possibility which has attracted some optimistic attention is the use of 'averaging' techniques. Here only the kinematic part of the diffracted intensities is extracted from the measured data. This procedure bears some principal similarities to the geometric theory, which enables the unit cell to be derived without solving the complete scattering problem.

After the unit cell of a surface structure has been derived by inspection of the LEED pattern, there remain the following unknown parameters for a complete structure interpretation:

(i) The distance of the surface layer to the next substrate layer.
(ii) The lateral shifts between the atoms of the surface layer and those of the substrate.
(iii) The positions of the atoms within the unit cell with respect to each other. (This question is only of relevance in cases where the unit cell contains more than one particle).

These problems may be tackled in various ways. The spacing and registry between the overlayer and the substrate may be derived using a number of different diffraction spots (h_1, h_2) of the substrate if the positions on the energy-scale of the kinematic Bragg maxima are known. In the kinematic theory the energy of the mth Bragg maximum is given by the interference condition for the direction normal to the surface:

$$(s_\perp^0 + s_\perp) \cdot \frac{d}{\lambda} - g_{h_1 h_2} \cdot d_0 = m \tag{9.59}$$

where $s_\perp^0$ and $s_\perp$ are the components normal to the surface of the unit vectors for the propagation of the incident and the diffracted beam, d is the distance between the two topmost layers, d_0

is the lateral shift of the origin of the first layer with respect to that of the second layer, and $g_{h_1 h_2} = h_1 a_1^* + h_2 a_2^*$ is a reciprocal two dimensional lattice vector of the substrate net. From equ. (9.59) we obtain in the case of normal incidence onto the (100) surface of a fcc crystal the following relation for the energies of the kinematic Bragg peaks [171]:

$$U_m(h_1, h_2) = \frac{(2\pi\hbar)^2}{8md^2} \left\{ \left[m + \frac{h_1+h_2}{2} \right]^2 + h_1^2 + h_2^2 + \frac{1}{4} \cdot \frac{h_1^2 + h_2^2}{\left[m + \frac{h_1+h_2}{2} \right]^2} \right\} - U_m^0 \qquad (9.60)$$

U_m^0 is the correction for the "inner potential". In the range of higher electron energies (≥ 100 eV) it should probably be possible to derive the energies of the kinematic Bragg maxima from an analysis of the "Bragg envelopes".

There is however another more important method for extracting the kinematic part from the I/U-curves (i. e. mainly the positions of the kinematic Bragg maxima) which is due to Lagally et al. [122]. The technique involves keeping the electron energy and the momentum transfer (i. e. $k_0 - k$) constant and measuring curves for a series of azimuth and/or scattering angles. The results are then used to compose average I/U-curves. The reasoning behind this procedure is that the dynamic scattering factors depend on k_0 and on k individually, whereas the kinematic part is only determined by the difference $k - k_0$ (see equ. 9.26). If $k - k_0$ is held constant while k and k_0 are varied then it is to be expected that the dynamic contributions are eliminated by averaging. Fig. 9.37 shows the result of such a treatment in the case of I/U-curves from Ni(111).

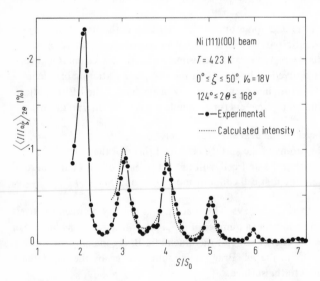

Ni (111)(00) beam
$T = 423$ K
$0° \leq \xi \leq 50°$, $V_0 = 18$ V
$124° \leq 2\theta \leq 168°$
—•—Experimental
········· Calculated intensity

Fig. 9.37. Experimental angular average for the intensities of the 00-beam from Ni(111) with correction for the inner potential (full line) and calculated kinematic data (dashed line). The scale of the abscissa is in units of $S/S_0 = 2d \sin \vartheta / \lambda$, where $2\vartheta = \pi - (\zeta + \zeta_*)$ is the angle between the directions of the incident and the diffracted beams. After Ngoc, Lagally and Webb [122c].

Obviously the averaged I/U-curve has strong kinematic character.

In principle it should be possible to derive the vertical distance and the registry between an adsorbed layer and the substrate surface in a similar manner. It must be realized however, that there is no theoretical proof for the assumption that all the multiple scattering contributions are eliminated by such a procedure [123]. Model calculations with s-wave scatterers by Duke and Smith [124] revealed that this method is only useful if a very large number (>1000) of experimental I/U-curves are averaged.

The remaining missing parameter in the complete LEED analysis is the evaluation of the positions of particles within the unit cell with respect to each other. As outlined previously there is no problem in those cases where for geometrical (i.e. particle size) or other reasons only a single particle is contained in the unit cell. Otherwise the structure factor of the unit cell is strongly dependent on the diffraction angle due to the phase differences of the scattered waves originating from the different atoms. If the kinematic theory is applicable then equ. (9.35) can be used to calculate the intensity distribution in the diffraction pattern at any fixed electron energy for an assumed structure model, provided that the angular variations of the individual atomic scattering factors are known. This is the principle underlying normal structure analysis by means of x-ray diffraction. Using a trial and error procedure a model is then selected which gives the best agreement between the calculated and observed intensity distribution in the diffraction pattern.

The application of this concept to two-dimensional surface layers is complicated by multiple scattering within the layer and by the existence of the substrate lattice. These problems can be tackled by the "energy-averaging" method as proposed by Tucker and Duke [125], who demonstrated that the existence of the substrate only results in fine structure in the I/U-curve of the allowed beams which arise from diffraction at the overlayer, and has only a minor influence on the *relative* intensities of the 'extra' spots. By integrating the intensities of individual 'extra' spots over an energy range ΔU it is hoped that multiple scattering effects are eliminated so that the kinematic theory and equ. (9.35) can be used. (Inadequate knowledge of the angular dependence of the atomic scattering factors is still a problem).

Duke and Tucker [125] tested this method with a c(2×8)-structure of oxygen adsorbed on Rh(100). The chosen energy range was between 70 and 250 eV. The relative intensities of the 'extra spots' were only estimated from their brightness on photographs from the diffraction patterns. The averaged intensities from the spots (h_1, h_2) were then compared with the squares of the corresponding kinematic structure factors

$$F_{h_1 h_2} = \sum_j <f_j> e^{2\pi i (h_1 x_j + h_2 y_j)} \tag{9.61}$$

x_j, y_j denotes the position of the jth atom within the unit cell and $\langle f_j \rangle$ is the average of the "effective" atomic scattering factor [126].

The intensities should be averaged over a not too small energy interval (50–100 eV) and for higher energies (>100 eV) where multiple scattering effects are not so dominating. If the scatterers are considered as being isotropic, then

$$f_j = \frac{e^{2i\delta j} - 1}{2 i k(E)} \tag{9.62}$$

Now f_j varies only slowly with $k(E)$, and Tucker and Duke [125] used constant values which were adjusted empirically as a parameter for the scattering properties of the individual atoms. An extension of this method would be to include an angular variation of the atomic scattering factors, which on the other hand can probably be determined experimentally using this theory if the surface structures are known [126]–[128].

Again there is no theoretical proof for the general validity of the assumption that the multiple scattering effects are eliminated by the averaging procedure [123]. A step in this direction was made by Tucker and Duke [125] who averaged intensities as calculated by a multiple scattering theory for a model of s-wave scatterers, which, however, implies only weak multiple scattering. In a recent calculation Duke and Laramore [128] used higher phase shifts but rather strong absorption, which again suppresses multiple scattering.

At present there are too few data available to give a sound prognosis whether in the future the analysis of surface structures will be obtained by these average techniques essentially by using empirical principles and parameters. The prospects for analyses of adsorbed layers are not very favourable since these methods are only generally applicable if the atoms exhibit very strong inelastic interactions with the scattered electrons [125]. This is a very severe requirement satisfied probably only by very few adsorbates (certainly not light elements like H and O), so that structure analysis of surfaces in the future will presumably mostly be based on the application of microscopic LEED theories as outlined in the preceding section and not on the just-described data-reduction techniques.

9.12. Structural aspects of clean surfaces

Some characteristic results of LEED studies on clean surfaces are discussed in this section and are used to demonstrate the information already available from LEED as well as the unsolved problems. Further results such as those concerning surface vibrations and adsorbate covered surfaces are dealt with in the last two chapters.

9.12.1. Atomic distances in surface layers

The interference properties of the LEED technique provide a means for determining the lattice constants parallel to the surface from the positions of the diffraction spots. If only one atom is contained in the unit cell then this yields the distance between the surface atoms.

At normal incidence of the primary beam the possible angles of diffraction φ are given by the wellknown relation

$$d_{\|} \cdot \sin \varphi = n \lambda = n \sqrt{\frac{150.4}{U}} \tag{9.63}$$

Here $d_{\|}$ denotes the distance between parallel rows in the surface, n is the order of diffraction and $U = U' - U_c$, the difference between the accelerating voltage U' and the contact potential U_c between the cathode and the target. The accuracy which can usually be achieved is some 3–5%. Within this accuracy the lattice constants in the surface as derived from the positions of 'normal' diffraction spots always agree with the bulk values. However this does not mean that the arrangement of the surface atoms is always the same as in the bulk, since a series of systems is known where a different surface periodicity gives rise to the appearance of 'extra' spots, as described below.

A determination of the surface lattice constant $d_{\|}$ with very high precision has been performed by Ekelund and Leygraf [129] for Si(111). From the 'normal' spots (there are also 'extra' spots!) they derived a value of $d_{\|} = 3.366 \pm 0.012$ Å. The corresponding bulk value $d_b = 3.32573$ Å seems to be slightly smaller, indicating probably a small lateral extension of the lattice in the surface region.

The distance $d_\perp$ between equivalent atomic layers perpendicular to the surface is connected with the formation of primary Bragg peaks in the I/U-curves. For normal incidence of the primary beam the position of the mth Bragg peak is determined within the kinematic approximation by the condition

$$m/d_\perp = 2 k_\perp \tag{9.64}$$

where $k_\perp$ is the component of the momentum normal to the surface. Since $k_\perp = k \cdot \sin \varphi$, and $k = 2\pi/\lambda$, it follows that the energies of the Bragg maxima U_m are given by

$$U_m = \frac{150.4 \ m^2}{4 d_\perp^2 \cdot \sin^2 \varphi} - \frac{U_0^2}{\sin^2 \varphi} \tag{9.65}$$

Note that the correction $U_0^2/\sin^2 \varphi$ has to be included due to the existence of the inner potential U_0. A plot of U_m versus m^2 should yield a straight line, and from this slope $d_\perp$ can be evaluated. Andersson and Kasemo [130] and Stern and Sinharoy [131] determined the values of $d_\perp$ for Ni(100), Cu(100) and W(110) from the intensity profiles of the 0,0-beams. Within the limits of accuracy the results agreed with the bulk data as derived from x-ray diffraction.

It would be wrong however to conclude from these results that no variation of the distance between atomic layers in the surface region exists, as is frequently suggested. The weakness lies in the assumption (which is based on the kinematic theory) that such information can be obtained unequivocally from a single intensity profile without a dynamic analysis. In fact the existence of small expansions or compressions of the distances between the two topmost layers are one example of the unresolved questions and mark the limitations of the present state of the theory:

For Al(110) surfaces a contraction of the topmost layers by about 10% was concluded from an analysis of the intensity profiles [132], whereas Laramore et al. [81] offer an alternate explanation in terms of the existence of a step distribution on the surface. Probably neither suggestion would be needed if a better value for the inner potential were used [17k]. Similar problems exist for the (100) planes of some of the alkali halides where different vertical shifts of the layers of cations and anions may possibly occur [133].

Lagally et al. [127] applied their averaging technique to intensity data from Ag(111) and Ni(111). The results do not allow a clear distinction to be made whether the surface layer is expanded or not. Recent dynamical calculations for Ni(100) and Ni(111) by Laramore [134] (which exhibit close agreement to the measured intensity data) indicate that the upper-layer spacing is the same as in the bulk to within 0.1 Å for both faces. This result is somewhat surprising in view of the large differences of the vibrational amplitudes of surface and bulk atoms as discussed in chapter 10.

9.12.2. Metals

LEED patterns from clean metal surfaces normally exhibit only diffraction spots which would arise from scattering at a plane in the bulk parallel to the surface, indicating that the periodicity in the surface is identical to that of the corresponding net plane in the bulk.

For some surfaces of the noble metals Ir, Pt, Au (which are neighbours in the periodic table of elements) in contrast to all other metal surfaces which have been investigated so far, characteristic additional diffraction spots have been observed which must be ascribed to special surface structures. Particular attention has been focussed on Au(100) [135–137] where two types of domains of a 5 × 1-structure rotated by 90° to each other have been observed. Furthermore these 'extra' spots are also split, so that a 20 × 5-structure describes the situation more exactly. Attempts have been made to explain this effect as due to the formation of a hexagonal surface layer consisting of Au atoms on a substrate with cubic symmetry[137] as drawn in fig. 9.38.

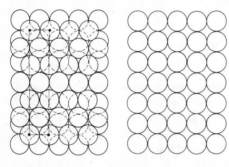

Reconstructed layer   Nonreconstructed layer

Fig. 9.38. Proposed model for the surface structure of clean Au(100). The reconstructed surface consists of a hexagonal layer on the nonreconstructed substrate. After Palmberg and Rhodin [137a].

A recent investigation using diffraction of high energy electrons (RHEED) [138] leads to the conclusion that the 5 × 20-structure is in reality built up of a hexagonal unit having one axis parallel to the Au(011) direction whereby $b_1^* = \frac{6}{5} a_1^*$ and $b_2^* = \frac{6}{5} (\frac{1}{2} a_1^* + \frac{1}{2} \sqrt{3} \, a_2^*)$. Thus in one direction the periodicity is 5, whereas the surface structure is incoherent in the other direction. However there are still some doubts as to whether this structure is exhibited by a clean surface or whether it is stabilized by small amounts of impurities. Palmberg and Rhodin [137] found this structure even after epitaxial growth of Au(100) in situ and didn't detect any surface impurities by means of AES. Fedak et al. [138] repeated the LEED and AES investigations and observed that chlorine is able to transform the 5 × 1-structure into the normal 1 × 1-structure.

Identical superstructures have been discovered in the case of Pt(100) [139] and Ir(100) [140]. Grant and Haas [141] concluded for Pt(100) that the 5 × 1-structure is caused by the presence of oxygen, which is probably situated below the surface layer and therefore cannot be detected by means of AES. Also with (110) surfaces of Ir [142], Pt [143] and Au [144] the formation of 1 × 2 superstructures has been observed without (at least in the case of Ir) any contaminants being found in the Auger spectra.

Goodman and Somorjai [145] performed LEED studies with different single crystal planes of Pb, Bi and Sn up to the melting point and during the melting process. The diffraction features persisted in all cases up to the melting points of the solids. This indicates that the surfaces remain ordered up to the melting temperature of the bulk and that obviously the surfaces play an important role in the melting process as nucleation or initiating centres. This observation is consistent with a series of other findings, where after proper handling of surfaces or other crystal imperfections considerable superheating above the melting point was possible.

There are several theories which emphasize the importance of the surface in the melting process; for example Stranski [146] considered melting as a process where a crystal dissolves in its own melt. Interesting studies on the melting of Pb which was epitaxially deposited on Cu surfaces were made by Henrion and Rhead [172] who showed that the melting temperature may be strongly influenced by the surface orientation.

9.12.3. Alloys

Two different metals may form either ordered or disordered alloys. The first possibility is similar to the formation of ordered adsorbed layers, insofar as the periodicity in the surface differs from that of the pure components. Several examples for the formation of ordered alloys and their observation by means of LEED are reported in the literature [147]. Cu and Au may form an ordered phase Cu Au, for which the lattice sites are alternately occupied by Cu or Au atoms. For the arrangement on the (100) surface therefore a $c(2 \times 2)$-structure results, which in fact has been observed by Palmberg and Rhodin [137b, c] with epitaxially grown layers.

For LEED studies with disordered alloy surfaces it can be shown [148] that besides the normal substrate spots the background intensity should be somewhat increased, depending on the difference between the atomic scattering factors of both kinds of atoms. LEED patterns from Cu/Ni [148] and Ag/Pd alloys [149] showed no significant increases of the background brightness which must be due to similar atomic scattering factors of the components.

9.12.4. Semiconductors and semimetals

Whereas with metals the formation of altered surface periodicities is restricted to a few exceptions, with semiconductors and semimetals this effect is more the rule than the exception [150]. A plausible explanation might possibly be found in the covalent character of the chemical bonds for these materials, which are much more localized than in the case of metals. If the

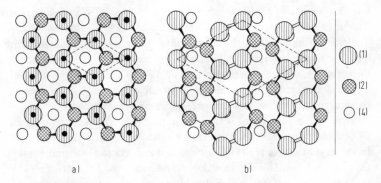

a)   b)

Fig. 9.39. Cleaved Ge(111) surface. a) Unreconstructed surface; b) Atomic arrangement corresponding to a 2×1-structure. Three domain orientations are possible on the substrate. (1) Top layer atoms, (2) second layer, (4) fourth layer. After Lander et al. [151].

complete coordination of the atoms is interrupted at the surface the occurrence of "dangling bonds" is to be expected, which may follow their tendency for saturation by periodic displacements of surface atoms.

The most detailed studies have been performed with Ge(111) and Si(111) surfaces. Both surfaces may be prepared by cleaving crystals at room temperature. In this case they exhibit in the LEED pattern 'extra' spots of a 2×1-structure [151]. A possible model for these structures is shown in fig. 9.39. The double periodicity is achieved by a pairwise mutual approach of surface atoms. However this picture illustrates only one of several proposed models, since a definite decision cannot be made without an intensity analysis.

It is important to note that these cleavage structures are only metastable states, for they transform irreversibly on annealing into new structures. The stable phase of a Ge(111) surface is a 2×8-structure [152]. Si(111) exhibits a 7×7-structure (fig. 9.40) [153] which has been

Fig. 9.40. LEED pattern for a clean Si(111) surface exhibiting a 7×7-structure.

the subject of numerous investigations and speculations. At temperatures above 800 °C a further (reversible) phase transformation from the 7×7- into a 1×1-structure takes place. The latter may be stabilized at room temperature by small amounts of adsorbed chlorine [165]. The transformation from the cleaved into the annealed surface structure has further consequences for the electronic and chemical properties of these surfaces. Henzler [154] was able to perform LEED measurements and field effect measurements on Ge(111) simultaneously and demonstrated the interrelation between electronic surface states and the geometrical configuration of the surface atoms. Similar observations were made with Si(111) [155]. The catalytic decomposition of water on Ge proceeds with an activation energy which is about 10 kcal/mole higher on annealed than on cleaved surfaces [156].

In the case of GaAs the (110) surface is the cleavage plane and contains equal numbers of Ga and As atoms. This surface exhibits an undistorted 1×1-structure [157], whereas the polar (111) and ($\overline{1}\overline{1}\overline{1}$)-surfaces (which should contain only atoms of one type) form 2×2- and 3×3-superstructures, respectively [158].

Similar observations have been made with Te [159], where the LEED pattern from the ($10\overline{1}0$) cleavage plane exhibits the ideal 1×1-structure, probably because no covalent bonds are broken in this surface. On the other hand a 2×1-structure has been observed with the (0001) surface.

LEED studies by Jona [160] with the most densely packed planes (0001), (01$\bar{1}$2) and (11$\bar{2}$0) of Sb and Bi revealed that the clean (0001) and (01$\bar{1}$2) surfaces have the ideal structure, whereas the (11$\bar{2}$0) surfaces undergo reconstruction in both cases. An explanation is given in terms of the tendency to lower the surface density of "dangling bonds" which seems to be reasonable, since the chemical bond in these materials has a high degree of covalency.

Highly interesting observations have been made by Palmberg et al. [161] with NiO. Antiferromagnetic domains are formed by periodically alternating spin orientations. Thus the periodicity of the spin orientation is twice the geometric atomic periodicity leading to the formation of half-order LEED spots. This example demonstrates the importance of the spin-spin interactions in the diffraction process.

Many other interesting LEED investigations have been made with semiconductors such as ZnO [166], materials like V_2O_5 [167], $BaTiO_3$ [168], $YMnO_3$ [169] and with graphite [170] and diamond [171].

9.12.5. Ionic crystals and insulators

LEED studies with ionic crystals have been performed mainly to check theoretical models with relatively simple systems [133]. Difficulties due to charging effects can usually be avoided by using thin samples which are heated to a few hundred degrees during the measurements. The (100) cleavage planes of the alkali halides were the main objects of investigation and in all cases only the 'normal' diffraction spots were detected [133a], [162]. The problems concerned with the interlayer spacing at the surface were already mentioned in section 9.12.1.

LEED investigations with insulator surfaces are also possible in principle, but only a few results have been obtained so far. Electronic charging of the surfaces may be the main problem. One possibility of eliminating these effects is to use intermittent pulses of high energy electrons in the primary beam. Due to the high secondary electron emission coefficient the surplus charge

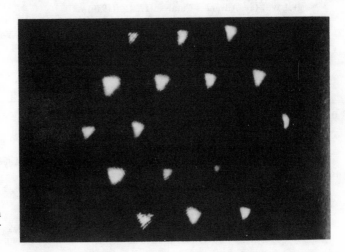

Fig. 9.41. LEED pattern of a cleaved mica surface. After Müller and Chang [164].

is removed from the surface. Some of the general problems with LEED studies at insulators have been discussed by Lang and Goldsztaub [163].

Interesting observations have been made by Müller and Chang [164] with cleaved surfaces of mica. Local electric fields appear to exist within the surface. Since three equivalent domain orientations are present this effect leads to the formation of LEED spots with a triangular shape as shown in fig. 9.41.

9.13. References

[1] L. de Broglie, Phil. Mag. *47*, 446 (1924).
[2] C.J. Davisson and C.H. Kunsman, Science *52*, 522 (1921).
[3] H.E. Farnsworth, Phys. Rev. *25*, 41 (1925).
[4] C.J. Davisson and C.H. Kunsman, Phys. Rev. *22*, 242 (1923).
[5] W. Elsasser, Naturw. *13*, 711 (1925).
[6] C.J. Davisson and L.H. Germer, Phys. Rev. *30*, 705 (1927).
[7] W. Ehrenberg, Phil. Mag. *18*, 878 (1934).
[8] E.J. Scheibner, L.H. Germer and C.D. Hartman, Rev. Sci. Instr. *31*, 112 (1960).
[9] J.J. Lander, F. Unterwald and J. Morrison, Rev. Sci. Instr. *33*, 784 (1962).
[10] C.W. Tucker, Appl. Phys. Lett. *2*, 34 (1962).
[11] W.T. Sproull, Rev. Sci. Instr. *4*, 193 (1933).
[12] R.L. Park and H.E. Farnsworth, Rev. Sci. Instr. *35*, 1592 (1964).
[13] C. Eckart, Proc. Nat. Acad. Amer. *13*, 460 (1927).
[14] H. Bethe, Naturwiss. *15*, 786 (1927); *16*, 333 (1928); Ann. Phys. *87*, 55 (1928).
[15] P.M. Morse, Phys. Rev. *35*, 1310 (1930).
[16] B.A. Stevens, Bibliography No. 141, Bell Telephone Lab. Techn. Inform. Libraries, Murray Hill, N.J., Oct. 1969.
[17] a) J.J. Lander, Progr. Solid State Chem. *2*, 26 (1965).
     b) V.F. Dvorjakin and A.Ju. Mityagin, Sov. Phys. Cryst. *12*, 982 (1967).
     c) H.E. Farnsworth in: Experimental Methods in Catalytic Research, (R.B. Anderson, ed.), Academic Press, New York 1968, p. 265.
     d) G.A. Somorjai, Ann. Rev. Phys. Chem. *19*, 251 (1968).
     e) P.J. Estrup and E.G. McRae, Surface Sci. *25*, 1 (1971).
     f) E. Bauer in: Techniques of Metals Research, (R.F. Bunshah, ed.), Wiley, New York 1969, Vol. 2, part 2, p. 559.
     g) J.W. May, Advances Catalysis *21*, 151 (1970).
     h) E.N. Sickafus and H.P. Bonzel in: Progress in Surface and Membrane Science, (Eds. Danielli, Rosenberg, Cadenhead), Vol. 4, p. 116, Academic Press, New York 1971.
     i) G.A. Somorjai and H.H. Farrell, Adv. Chem. Phys. *20*, 215 (1971).
     j) G.E. Laramore, J. Vac. Sci. Techn. *9*, 625 (1972).
     k) F. Forstmann in: Festkörperprobleme, Vol. 13 (H.J. Queisser, ed.), Vieweg Verlag, Braunschweig 1973.
     l) C.B. Duke, in: LEED – Surface Structure of Solids, Vol. 2 (Ed. L. Laznicka), Union of Czechoslovak Mathematicians and Physicists, Prague 1972, p. 125
     m) J.B. Pendry, ibid. p. 305.
     n) M.G. Lagally and M.B. Webb, in: Solid State Physics, Vol. 28 (Ehrenreich, Seitz and Turnbull, eds.), Academic Press, New York 1973, p. 302.
     o) C.B. Duke, in: Electron Emission Spectroscopy, (Eds. W. Dekeyser et al.), D. Reidel, Dordrecht 1973, p. 1.
     p) J.B. Pendry: Low energy electron diffraction. Academic Press, New York 1974.
     q) C.B. Duke, Adv. Chem. Phys. *27*, 1 (1974).
[18] E.A. Wood, J. appl. Phys. *35*, 1306 (1964).
[19] R.L. Park and H.H. Madden, Surface Sci. *11*, 188 (1968).
[20] C.W. Tucker, J. appl. Phys. *35*, 1897 (1964); *37*, 528, 3013 (1966).

[21] A. Fingerland, Surface Sci. *32*, 729 (1972).

[22] B. Lang, R.W. Joyner and G.A. Somorjai, Surface Sci. *30*, 440 (1972).

[23] R.N. Lee, Rev. Sci. Instr. *39*, 1306 (1968).

[24] J.C. Tracy, Rev. Sci. Instr. *39*, 1300 (1968).

[25] a) J. Morrison, Rev. Sci. Instr. *37*, 1263 (1966).
b) P.W. Palmberg, Rev. Sci. Instr. *38*, 834 (1967).

[26] C.W. Caldwell, Rev. Sci. Instr. *36*, 1500 (1965).

[27] P.W. Palmberg, Appl. Phys. Lett. *19*, 183 (1969).

[28] K. Fujiwara, K. Hayakawa and S. Miyaka, Jap. J. Appl. Phys. *5*, 295 (1966).

[29] H.E. Farnsworth, Phys. Rev. *34*, 679 (1929).

[30] R.L. Park and H.E. Farnsworth, Rev. Sci. Instr. *35*, 1592 (1964).

[31] G. Ertl and J. Koch, Z. phys. Chem. N.F. (Frankfurt) *69*, 323 (1970).

[32] L. Brillouin: Wave Propagation in Periodic Structures. McGraw Hill, New York 1946.

[33] R.W. James: The optical principles of the diffraction of x-rays. Bell, London 1962.

[34] G. Ertl and P. Rau, Surface Sci. *15*, 443 (1969).

[35] T.W. Haas, A.G. Jackson and M.P. Hooker, J. Chem. Phys. *46*, 3025 (1967).

[36] J.J. Lander, G.W. Gobeli and J. Morrison, J. Appl. Phys. *34*, 2298 (1963).

[37] a) G. Ertl, Surface Sci. *6*, 208 (1967).
b) G.W. Simmons, D.F. Mitchell and K.R. Lawless, Surface Sci. *8*, 130 (1967).

[38] G. Ertl and J. Koch in: Adsorption-Desorption Phenomena. (F. Ricca, ed.), Academic Press, New York 1972, p. 345.

[39] E. Bauer, Surface Sci. *7*, 351 (1967).

[40] a) N.J. Taylor, Surface Sci. *4*, 161 (1966).
b) E. Bauer, Phys. Rev. *123*, 1207 (1961).
c) C.W. Tucker, J. appl. Phys. *35*, 1897 (1964).

[41] P.W. Palmberg, Surface Sci. *25*, 598 (1971).

[42] F. Jona, R.F. Lever and J.B. Gunn, Surface Sci. *9*, 468 (1968).

[43] P.W. Palmberg and T.N. Rhodin, J. Chem. Phys. *49*, 147 (1968).

[44] D.G. Fedak and N.A. Gjostein, Surface Sci. *8*, 77 (1967).

[45] C. Kittel: Introduction to Solid State Physics. John Wiley, New York 1968.

[46] M.G. Lagally, T.C. Ngoc and M.B. Webb, Phys. Rev. Letters *26*, 1557 (1971).

[47] M.G. Lagally, T.C. Ngoc and M.B. Webb, J. Vac. Sci. Techn. *9*, 645 (1972).

[48] A. Guinier: X-Ray Diffraction in Crystals. Freeman, San Francisco 1963.

[49] A. Chutjian, Phys. Lett. *A24*, 615 (1967).

[50] K. Christmann, G. Ertl and O. Schober, Surface Sci. *40*, 61 (1973).

[51] F. Jona, Surface Sci. *8*, 478 (1967).

[52] R. Hosemann and S.N. Bagchi: Direct analysis of matter by diffraction. North Holland, Amsterdam 1962.

[53] R.L. Park, J.E. Houston and D.G. Schreiner, Rev. Sci. Instr. *42*, 60 (1971).

[54] R.D. Heidenreich: Fundamentals of transmission electron microscopy. Interscience, New York 1964.

[55] R.L. Park in: The structure and chemistry of solid surfaces. (G.A. Somorjai, ed.) Wiley, New York 1969, p. 28-1.

[56] G. Ertl, in: Molecular processes on solid surfaces. (Drauglis, Gretz, Jaffee, eds.), McGraw Hill, New York 1969, p. 147.

[57] R.L. Park and J.E. Houston, Surface Sci. *18*, 213 (1969).

[58] J.E. Houston and R.L. Park, Surface Sci. *21*, 209 (1970).

[59] L.H. Germer, J.W. May and R.J. Szostak, Surface Sci. *7*, 430 (1967).

[60] P.J. Estrup and J. Andersson, J. Chem. Phys. *45*, 2254 (1966).

[61] K. Molière and R. Portele, in: The structure and chemistry of solid surfaces. (G.A. Somorjai, ed.), Wiley, New York 1969, p. 69-1.

[62] K. Fujiwara, J. Phys. Soc. Japan *12*, 7 (1957).

[63] R. Heckingbottom, Surface Sci. *27*, 370 (1971).

[64] G. Ertl and J. Küppers, Surface Sci. *21*, 61 (1970).

[65] a) R. Gerlach and T.N. Rhodin, Surface Sci. *10*, 446 (1968); *17*, 32 (1969).
b) J. Domange and J. Oudar, Surface Sci. *11*, 124 (1968).
c) J.C. Tracy and P.W. Palmberg, J. Chem. Phys. *51*, 4852 (1969).

[66] See for example: A. U. MacRae, Surface Sci. *4*, 247 (1966).
[67] H. Mykura, in: Molecular Processes on Solid Surfaces. (Drauglis, Gretz, Jaffee, eds.), McGraw Hill, New York 1969, p. 129.
[68] A. U. MacRae, Appl. Phys. Lett. *2*, 88 (1963).
[69] M. Boudart and D. F. Ollis, in: The structure and chemistry of solid surfaces. (G. A. Somorjai, ed.) Wiley, New York 1969, p. 63-1
[70] C. W. Tucker, J. Appl. Phys. *38*, 1988 (1967).
[71] J. C. Tracy and J. M. Blakely, Surface Sci. *13*, 313 (1969).
[72] a) N. J. Taylor, Surface Sci. *2*, 544 (1964).
     b) C. C. Chang and L. H. Germer, Surface Sci. *8*, 115 (1967).
     c) J. Anderson and W. E. Danforth, J. Franklin Inst. *279*, 160 (1965).
     d) L. H. Germer and J. W. May, Surface Sci. *4*, 452 (1966).
[73] J. T. Grant and T. W. Haas, Surface Sci. *17*, 484 (1969).
[74] J. J. Lander and J. Morrison, Surface Sci. *2*, 553 (1964).
[75] W. P. Ellis and R. L. Schwoebel, Surface Sci. *11*, 82 (1968).
[76] M. Henzler, Surface Sci. *19*, 159 (1970).
[77] T. W. Haas, Surface Sci. *5*, 345 (1966).
[78] L. G. Feinstein and M. S. Macrakis, Surface Sci. *18*, 277 (1969).
[79] G. E. Rhead and J. Perdereau, Surface Sci. *24*, 555 (1971).
[80] J. E. Houston and R. L. Park, Surface Sci. *26*, 269 (1971).
[81] G. E. Laramore, J. E. Houston and R. L. Park, J. Vac. Sci. Techn. *10*, 196 (1973).
[82] C. B. Duke and A. Liebsch, Phys. Rev. *B9*, 1126; 1150 (1974).
[83] W. P. Ellis and B. D. Campbell, in Proc. of LEED Symposium, Am. Cryst. Ass., Tucson 1968; Polycrystal Book Service, Pittsburgh.
[84] D. G. Fedak, T. E. Fischer and W. D. Robertson, J. appl. Phys. *39*, 5658 (1968).
[85] W. P. Ellis, in: Optical transforms. (H. S. Lipson, ed.) Academic Press, New York 1972.
[86] N. F. Mott and H. S. W. Massey: The theory of atomic collisions. Oxford University Press, London 1949.
[87] A. Ignatjevs, J. B. Pendry and T. N. Rhodin, Phys. Rev. Lett. *26*, 189 (1971).
[88] E. G. McRae, J. Chem. Phys. *45*, 3258 (1966).
[89] a) E. G. McRae, Surface Sci. *8*, 14 (1967).
     b) R. L. Gerlach and T. N. Rhodin, Surface Sci. *8*, 1 (1967).
[90] a) L. H. Germer, Physics Today *17*, 19 (1964).
     b) L. H. Germer, Surface Sci. *5*, 147 (1966).
[91] a) E. Bauer, in: Adsorption et Croissance Cristalline. CNRS, Paris 1965, p. 20.
     b) E. Bauer, Surface Sci. *5*, 152 (1966).
     c) C. W. Tucker, Surface Sci. *26*, 311 (1971).
[92] P. J. Jennings and E. G. McRae, Surface Sci. *23*, 363 (1971).
[93] W. A. Harrison: Pseudopotentials in the Theory of Metals. W. A. Benjamin, New York 1966.
[94] a) C. B. Duke and G. A. Laramore, Phys. Rev. *B2*, 4765; 4783 (1970).
     b) C. B. Duke, D. L. Smith and B. W. Holland, Phys. Rev. *B5*, 3358 (1972).
[95] a) J. B. Pendry, J. Phys. *C2*, 2273; 2283 (1969).
     b) V. Hoffstein and D. Boudreaux, Phys. Rev. Lett. *25*, 512 (1970).
     c) J. A. Strozier and R. O. Jones, Phys. Rev. Lett. *25*, 516 (1970).
[96] a) J. B. Pendry, J. Phys. *C4*, 2501; 3095 (1971).
     b) J. A. Strozier and R. O. Jones, Phys. Rev. *B3*, 3228 (1971).
[97] F. Forstmann, W. Berndt and P. Büttner, Phys. Rev. Lett. *30*, 17 (1973).
[98] a) D. W. Jepsen, P. M. Marcus and F. Jona, Phys. Rev. *B5*, 3933 (1972), *B6*, 3684 (1972).
     b) G. E. Laramore and C. B. Duke, Phys. Rev. *B5*, 267 (1972).
[99] a) J. A. Appelbaum and D. R. Hamann, Phys. Rev. *B6*, 2166 (1972).
     b) N. D. Lang, in: Solid State Physics Vol. 28 (Ehrenreich, Seitz and Turnbull, eds.), Academic Press, New York 1973, p. 225.
[100] S. Andersson and J. B. Pendry, J. Phys. *C5*, L41 (1972), *C6*, 601 (1973).
[101] D. W. Jepsen, P. M. Marcus and F. Jona, Phys. Rev. *B5*, 933 (1972).
[102] C. B. Duke and C. W. Tucker, Surface Sci. *15*, 231 (1969).
[103] C. G. Darwin, Phil. Mag. *27*, 675 (1914).

[104] E.G. McRae, Surface Sci. *11*, 429 (1969); *25*, 491 (1971).

[105] G. Gafner, in: The structure and chemistry of solid surfaces. (G.A. Somorjai, ed.), Wiley, New York 1969, p. 2-1.

[106] M. Lax, Rev. Mod. Phys. *23*, 287 (1951); Phys. Rev. *85*, 621 (1952).

[107] E.G. McRae and C.W. Caldwell, Surface Sci. *7*, 41 (1967).

[108] K. Kambe, Z. Naturf. *22a*, 322; 422 (1967).

[109] J.C. Slater, Phys. Rev. *31*, 140 (1937).

[110] a) C.B. Duke and C.W. Tucker, Surface Sci. *24*, 31 (1971).
b) C.B. Duke, J.R. Anderson and C.W. Tucker, Surface Sci. *19*, 117 (1970).
c) C.B. Duke and C.W. Tucker, J. Vac. Sci. Techn. *8*, 5 (1971).
d) C.W. Tucker and C.B. Duke, Surface Sci. *23*, 411 (1970).
e) C.B. Duke and C.W. Tucker, Phys. Rev. *B3*, 3561 (1971).

[111] J.L. Beeby, J. Phys. *C1*, 82 (1968).

[112] C.M.K. Watts, J. Phys. *C2*, 966 (1969); *C1*, 1237 (1968).

[113] a) P.M. Marcus and D.W. Jepsen, Phys. Rev. Lett. *20*, 925 (1968).
b) P.M. Marcus, D.W. Jepsen and F. Jona, Surface Sci. *17*, 442 (1969).

[114] J.B. Pendry, J. Phys. *C4*, 2501 (1971).

[115] H. Bethe, Ann. Phys. *87*, 55 (1928).

[116] D.S. Boudreaux and V. Heine, Surface Sci. *8*, 426 (1967).

[117] a) R.O. Jones and J.A. Strozier, Phys. Rev. Lett. *22*, 1186 (1969).
b) J.B. Pendry, J. Phys. *C2*, 2273 (1969).
c) V. Hoffstein and D.S. Boudreaux, Phys. Rev. Lett. *25*, 512 (1970).

[118] G. Capart, Surface Sci. *13*, 361 (1969); *26*, 429 (1971).

[119] J.B. Pendry, J. Phys. *C4*, 3095 (1971).

[120] R.H. Tait, Y. Tong and T.N. Rhodin, Phys. Rev. Lett. *28*, 553 (1972).

[121] M.G. Lagally, Z. Naturforschung *25a*, 1567 (1970).

[122] M.G. Lagally, T.C. Ngoc and M.B. Webb,
a) Phys. Rev. Lett. *26*, 1557 (1971);
b) J. Vac. Sci. Techn. *9*, 645 (1972);
c) Surface Sci. *35*, 117 (1973).

[123] J.B. Pendry, J. Phys. *C5*, 2567 (1972).

[124] C.B. Duke and D.L. Smith, Phys. Rev. *B5*, 4730 (1972).

[125] C.W. Tucker and C.B. Duke, Surface Sci. *23*, 411 (1970); *29*, 237 (1972).

[126] M.G. Lagally and M.B. Webb, Phys. Rev. Lett. *21*, 1388 (1968).

[127] M.G. Lagally and M.B. Webb, in: The structure and chemistry of solid surfaces. (G.A. Somorjai, ed.) Wiley, New York 1969, p. 20-1.

[128] C.B. Duke and G.E. Laramore, Surface Sci. *30*, 659 (1972).

[129] S. Ekelund and C. Leygraf, Proc. Congr. on Vacuum Instruments and Methods in Surface Science, Guildford (U.K.), 1972.

[130] S. Anderson and B. Kasemo, Surface Sci. *25*, 273 (1971).

[131] R.M. Stern and S. Sinharoy, Surface Sci. *33*, 131 (1972).

[132] a) G.E. Laramore, C.B. Duke, A. Bagchi and A.B. Kunz, Phys. Rev. *B4*, 2058 (1971).
b) G.E. Laramore and C.B. Duke, Phys. Rev. *B5*, 267 (1972).

[133] a) E.G. McRae and C.W. Caldwell, Surface Sci. *2*, 509 (1964).
b) B.W. Holland, R.W. Hannum and A.M. Gibbons, Surface Sci. *25*, 561 (1971).
c) G.E. Laramore and A.C. Switendick, Phys. Rev. *B7*, 3615 (1973).

[134] G.E. Laramore, Phys. Rev. *B8*, 515 (1973).

[135] G.A. Somorjai, Surface Sci. *8*, 98 (1967).

[136] a) D.G. Fedak and N.A. Gjostein, Phys. Rev. Lett. *16*, 171 (1966).
b) D.G. Fedak and N.A. Gjostein, Surface Sci. *8*, 77 (1967).

[137] a) P.W. Palmberg and T.N. Rhodin, Phys. Rev. *161*, 586 (1967).
b) P.W. Palmberg and T.N. Rhodin, J. appl. Phys. *39*, 2425 (1968).
c) P.W. Palmberg and T.N. Rhodin, J. Chem. Phys. *49*, 134 (1968).

[138] D.G. Fedak, J.V. Florio and W.D. Robertson, in: The structure and chemistry of solid surfaces. (G.A. Somorjai, ed.) Wiley, New York 1969, p. 74-1.

[139] H.B. Lyon and G.A. Somorjai, J. Chem. Phys. *46*, 2539 (1967).

[140] a) J.T. Grant, Surface Sci. *18*, 288 (1969).
  b) A. Ignatiev, A.V. Jones and T.N. Rhodin, Surface Sci. *30*, 573 (1972).
[141] J.T. Grant and T.W. Haas, Surface Sci. *18*, 457 (1969).
[142] K. Christmann and G. Ertl, Z. Naturforsch. *28a*, 1144 (1973).
[143] H.P. Bonzel and R. Ku, J. Vac. Techn. *9*, 663 (1972).
[144] J. Oudar, personal communication.
[145] R.M. Goodman and G.A. Somorjai, J. Chem. Phys. *52*, 6325 (1970).
[146] I.N. Stranski, W. Gans and H. Rau, Ber. Bunsengesellsch. *67*, 965 (1963).
[147] a) N.J. Taylor, Surface Sci. *4*, 161 (1966).
  b) L.G. Feinstein and E. Blanc, Surface Sci. *18*, 350 (1969).
  c) P.W. Palmberg and T.N. Rhodin, J. Chem. Phys. *49*, 134 (1968).
[148] G. Ertl and J. Küppers, Surface Sci. *24*, 104 (1971); J. Vac. Sci. Techn. *9*, 829 (1972).
[149] K. Christmann and G. Ertl, Surface Sci. *33*, 254 (1972).
[150] See for example: G. Ertl and H. Gerischer, in: Physical Chemistry. An advanced treatise. (Eds. Eyring, Henderson, Jost) Academic Press, New York 1970, Vol. 10, p. 371.
[151] J.J. Lander, G.W. Gobeli and J. Morrison, J. appl. Phys. *34*, 2298 (1963).
[152] P.W. Palmberg and W.T. Peria, Surface Sci. *6*, 57 (1967).
[153] J.J. Lander and J. Morrison, J. Chem. Phys. *37*, 329 (1962).
[154] M. Henzler, J. appl. Phys. *40*, 3758 (1969).
[155] F. Bäuerle, W. Mönch and M. Henzler, J. Appl. Phys. *43*, 3917 (1972).
[156] G. Ertl and T. Giovanelli, Ber. Bunsenges. *72*, 74 (1968).
[157] A.U. MacRae and G.W. Gobeli, J. Appl. Phys. *35*, 1629 (1964).
[158] A.U. MacRae, Surface Sci. *4*, 247 (1966).
[159] S. Andersson, D. Andersson and I. Marklund, Surface Sci. *12*, 284 (1968).
[160] F. Jona, Surface Sci. *8*, 57 (1967).
[161] P.W. Palmberg, R.E. de Wames and L.A. Vredevoe, Phys. Rev. Lett. *21*, 682 (1968); J. Appl. Phys. *40*, 1158 (1969).
[162] a) I. Marklund and S. Andersson, Surface Sci. *5*, 197 (1966).
  b) T.E. Gallon, I.G. Higginbotham, M. Prutton and H. Tokotaka, Surface Sci. *21*, 224 (1970).
[163] B. Lang and S. Goldsztaub, Surface Sci. *32*, 473 (1972).
[164] K. Müller and C.C. Chang, Surface Sci. *9*, 455 (1968).
[165] J.V. Florio and W.D. Robertson, Surface Sci. *24*, 173 (1971).
[166] J.D. Levine, A. Willis, W.R. Bottoms and P. Mark, Surface Sci. *29*, 144 (1972).
[167] L. Fiermans and J. Vennik, Surface Sci. *18*, 317 (1969).
[168] D. Aberdam and C. Gaubert, Surface Sci. *27*, 559 (1971).
[169] D. Aberdam, G. Bouchet, P. Ducros, J. Daval, and G. Grunberg, Surface Sci. *14*, 121 (1969).
[170] J.J. Lander and J. Morrison, Surface Sci. *6*, 1 (1967).
[171] a) J.B. Marsh and H.E. Farnsworth, Surface Sci. *1*, 3 (1964).
  b) J.J. Lander and J. Morrison, Surface Sci. *4*, 241 (1966).
[172] J. Henrion and G.E. Rhead, Surface Sci. *29*, 20 (1972).
[173] S. Andersson and J.B. Pendry, J. Phys. *C6*, 601 (1973).
[174] C.B. Duke, N.O. Lipari and U. Landman, Phys. Rev. *B8*, 2454 (1973).
[175] M.R. Martin and G.A. Somorjai, Phys. Rev. *B7*, 3607 (1973).
[176] S.Y. Tong, T.N. Rhodin and R.H. Tait, Phys. Rev. *B8*, 430 (1973).
[177] J.E. Demuth, D.W. Jepsen and P.M. Marcus, Phys. Rev. Lett. *31*, 540 (1973).
[178] S. Andersson, B. Kasemo, J.B. Pendry and M.A. Van Hove, Phys. Rev. Lett. *31*, 595 (1973).
[179] C.B. Duke, N.O. Lipari and G.E. Laramore, Nuovo Cimento (in press).
[180] C.B. Duke, N.O. Lipari, G.E. Laramore and J.B. Theeten, Solid State Comm. *13*, 579 (1973).

10. Vibrations at Surfaces

10.1. Introduction

In the preceding chapters the solid and its surfaces were always treated as a rigid lattice, i.e. the positions of the atomic nuclei were assumed not to change with time. In fact in each solid the atoms are oscillating around their equilibrium positions due to their thermal energy. Since the situation at the surface is generally different from the bulk, the atomic motions may also differ from those in the bulk and cause a series of new effects.

In general any elastic distortion at the surface will give rise to a surface wave. These are excitations which can be described by wave vectors $k_\parallel$ for the propagation parallel to the surface and whose amplitudes decay with increasing depth into the bulk. The following three types of lattice surface waves may be distinguished [1]:

a) optical surface modes in ionic crystals having long wavelength;

b) long wavelength elastic surface waves, and

c) short-wavelength acoustic and optical surface phonons controlled by the short range force constants. A phonon characterizes the energy quantum of a lattice vibration. It can be considered as a quasi-particle with energy $\hbar\omega_p$, if ω_p is the frequency of the vibration.

Surface phonons may influence such properties as the specific heat, the surface free energy, the mean square displacement of surface atoms, thermal expansion etc. These phonons may be localized or propagate parallel to the surface.

A series of theoretical investigations concerned with such problems has been performed in the past mainly based on atomic force constant models. Detailed calculations for thin crystals consisting only of a few atomic layers have been made by Allen et al. [2] using a Lennard-Jones 6–12 potential. One of the results is that a tendency for an increase of the lattice spacing near the surface is predicted. Another consequence should be a T^2 law for the specific heat in the surface instead of the usual T^3 behaviour in the bulk [3].

Thermal expansion of a lattice is caused by anharmonic potentials for the vibrations of the lattice atoms. Variations of the lattice constant normal to the surface should be detected from elastic LEED experiments by shifts of the Bragg peaks on the energy scale of I/U-curves.

The main effect of the interaction between a vibrating lattice and slow electrons consists in variations of their scattering properties. The following two types of inelastic scattering must be distinguished [4]:

a) Interactions with spatially extended dipole fields (dipole scattering). If a surface wave or vibration is associated with a changing electric dipole field then a resonance-like interaction with a moving electron will take place if the component of the electron velocity in the direction of the alternating electric field is equal to its phase velocity. Such effects may lead to characteristic energy losses in the range of about 10–100 meV and have been discovered by Ibach [5] using electron spectrometers with very high resolution. A quantum mechanical treatment by Lucas and Sunjic [6] revealed that the loss probability increases with decreasing electron energy so that the primary energies used are best in the range of a few eV.

In the usual LEED system these losses cannot be separated from the elastic electrons and therefore form the "quasielastic" part.

b) Scattering from vibrating localized atomic potentials (impact scattering). This effect comprises the influence of temperature on the LEED intensities.

This influence is twofold:

α) The intensities of the diffracted beams are reduced. This effect is governed by the Debye-Waller factor (as in x-ray diffraction or Mößbauer spectroscopy), which depends on the mean square displacement of atoms in the direction of the scattering vector Δk.

β) The diffuse background intensity increases. This effect is caused by the thermal diffuse scattering and depends on the amplitudes of the normal vibrations of the system. In contrast to dipole scattering the probability for impact scattering, i.e. the effect of temperature in LEED observations, increases with the electron energy.

If particles are adsorbed on a clean surface further effects appear which are of major interest in surface chemistry. If the mass of the adsorbed particle differs sufficiently from that of the atoms of the solid (or if the binding energy is high enough) new localized mode frequencies of surface vibrations may occur which characterize the surface bond. A theoretical treatment by Dobrzynski et al. [7] revealed that these new frequencies should be above the band-frequencies (surface phonons) of the crystal. In fact such vibrations attributable to adsorbed particles were first detected experimentally by Propst and Piper [8] using high resolution loss spectroscopy. If the adsorbed particles are molecules, then the frequencies of their normal modes will be altered in a characteristic manner owing to the formation of the adsorption bond. Investigations of this type with ill-defined surfaces have been performed using infrared spectroscopy for a number of years. In this area some remarkable experimental progress has recently been made, which now enables also the investigation of adsorbed layers on well-defined single crystal surfaces. Although this would be outside the scope of this book, some of these recent IR-results are included in this chapter in order to demonstrate the present state of the art.

A further consequence of the creation of new surface vibrations by adsorbed particles may be a variation in the surface entropy, possibly giving rise to modifications of the pre-exponential factor in desorption kinetics, or in changes of the energy transfer between colliding particles and a crystal surface, which should decrease with increasing surface phonon frequencies [7b]. Experimental evidence for these effects is rather thin on the ground, however.

10.2. The vibrating lattice/temperature effects in LEED

10.2.1. Variations of the LEED-intensities

In a LEED experiment with a clean surface the influence of temperature becomes evident in that the intensities of the diffraction spots gradually decrease with increasing temperature and finally disappear in the background which becomes continuously brighter. Usually a plot of the logarithm of the intensity of a chosen spot at fixed energy vs. T yields a straight line whose slope depends on the electron energy.

In order to understand this effect the positions of the atoms may be no longer considered as being fixed but as changing their location with time due to thermal vibrations, i.e.

$$r_j(t) = {}^{\circ}r_j + u_j(t) \tag{10.1}$$

where ${}^{\circ}r_j$ denotes the equilibrium position and $u_j(t)$ the time dependent motion.

In the kinematic approximation this case corresponds directly to the temperature effect on the intensities in x-ray diffraction which is discussed in detail in the textbooks [9]. According to equ. (9.34) the structure factor F is given by

$$F = \sum_{j=1}^{s} f_j \cdot e^{i(k-k_0) \cdot r_j} = \sum_j f_j \, e^{-i\Delta k \cdot r_j} \tag{9.34}$$

where $\Delta k = k_0 - k$ is determined by the geometry and the electron energy.

The intensity of the elastically diffracted beam is proportional to the square of the structure factor. Insertion of the time dependent expression (10.1) for the vectors r_j leads to

$$F = \sum_j f_j \exp\left[-i\Delta k \cdot ({}^\circ r_j + u_j(t)]\right.$$

$$= \sum_j f_j \exp\left(-i\Delta k \cdot {}^\circ r_j\right) \cdot \exp\left[-i\Delta k \cdot u_j(t)\right] \tag{10.2}$$

The interesting quantity is the time averaged mean value $\langle F \rangle$. If one assumes that the motions of the atoms are independent of each other the averaged intensity is given by

$$I \propto |\langle F \rangle|^2 = \sum_j f_j \exp\left(-i\Delta k \cdot {}^\circ r_j\right)^2 \cdot \langle \exp\left[-i\Delta k \cdot u(t)\right)\rangle^2 \tag{10.3}$$

The first term in the product represents the intensity I_0 which would arise from a rigid lattice. Therefore

$$I = I_0 \cdot \langle \exp\left[-i\Delta k \cdot u(t)\right]\rangle^2 = I_0 \exp\left[(\Delta k)^2 \cdot \langle u^2(t)\rangle\right] \tag{10.4}$$

If the vibrations of the lattice atoms are characterized by a frequency spectrum such as used in Debye's theory for the specific heat of solids (Debye spectrum) then equ. (10.4) may be transformed into

$$I = I_0 \, e^{-2W} \tag{10.5}$$

where

$$2W = \frac{12h^2}{mk_B} \cdot \left(\frac{\cos\varphi}{\lambda}\right)^2 \cdot \frac{T}{\theta_D^2} \tag{10.6}$$

with m the mass of the atoms, k_B Boltzmann's constant, and φ the scattering angle of electrons with wavelength λ. θ_D is the characteristic Debye temperature. The mean square displacement $\langle u^2 \rangle$ is related with θ_D by the equation

$$\langle u^2 \rangle = \frac{3h^2}{4\pi^2 mk_B} \cdot \frac{T}{\theta_D^2} \tag{10.7}$$

The term e^{-2W} is called the Debye-Waller factor and causes an exponential decrease of the elastically scattered intensity with increasing temperature since $W \propto T$.

It can also be seen from equ. (10.4) that only that component of $u(t)$ which is parallel to Δk contributes to the damping of the intensity. Beams with low values of Δk suffer less under the temperature effect than those with large values of Δk.

A series of experimental investigations revealed the validity of a linear relation between log I and T, thus in principle enabling the determination of the Debye temperature θ_D or of $\langle u^2 \rangle$. The results have been reviewed by Tabor et al. [25].

Already in 1962 Germer and MacRae [10] were able to demonstrate that in the case of Ni(110) the surface Debye-temperatures as derived from LEED data are much smaller than the bulk values as evaluated from x-ray diffraction data. A generally observed trend is that with increasing electron energies the derived effective Debye temperatures gradually approach the bulk value. As an example data for a Ni(100) surface [11] are reproduced in fig. 10.1. Since

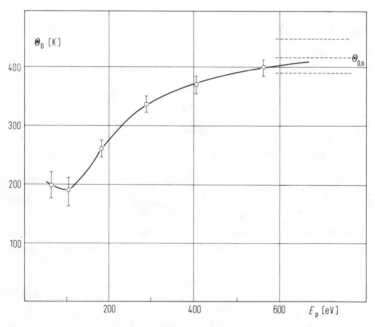

Fig. 10.1. Variation of the effective Debye temperature θ_D for Ni(100) with the energy of the electrons used for the determination of θ_D [11].

the depth of penetration and therefore the detected volume increases with increasing electron energy it has to be concluded that for the topmost atomic layer θ_D is much smaller (and $\langle u^2 \rangle$ much larger) than in the bulk.

Measurements by MacRae [12] with the 0,0-beam scattered at a Ni(110) surface revealed that θ_D becomes smaller if the primary electrons have nearly normal incidence than for grazing incidence (fig. 10.2). It follows from equ. (10.4) that only the component of u parallel to Δk is of importance. Therefore it is concluded that $\langle u^2 \rangle$ in the direction of the surface normal is larger than the mean square displacement parallel to the surface. Since the Ni(110) surface is anisotropic two different values for $\theta_{D,\parallel}$ have been derived. Detailed measurements with

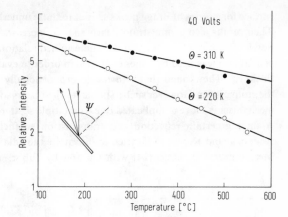

Fig. 10.2. Intensity of electrons elastically back-scattered from a Ni(110) surface close to the $[1\bar{1}0]$ direction with $\psi = 0°$ (open circles) and $\psi = 70°$ (filled circles) as a function of temperature. Electron energy = 40 Volts. After MacRae [12].

Ag surfaces by Somorjai et al. [13] revealed on the other hand no significant differences between $\theta_{D, \parallel}$ and $\theta_{D, \perp}$.

The dynamic aspects of solid surfaces have been treated theoretically by a number of authors. The earlier papers were reviewed by Maradudin [14].

Recently, Wallis et al. [15] performed calculations for fcc crystals assuming central forces between nearest-neighbour atoms. The force constants were derived from the elastic constants of the bulk materials. The results for (100) surfaces of the investigated metals (Cu, Ag, Au, Al, Pb, Ni, Pd, Pt) indicate that in all cases

$$\langle u^2 \rangle \perp \text{surface} \approx 2.0 \cdot \langle u^2 \rangle \text{bulk}$$

$$\langle u^2 \rangle \parallel \text{surface} \approx 1.2 - 1.5 \cdot \langle u^2 \rangle \text{bulk}$$

For a Ni(110) surface theoretical values [16] can be compared with the experimental data of MacRae [12]:

Surface direction	[110]	[1̄10]	[001]	bulk
Theory	0.8	0.64	0.86	0.40
Experiment	1.41	0.63	1.41	0.45

(The data are given in units of kT/α, where $\alpha = 3.79 \cdot 10^4$ dyn/cm is the bulk force constant for nickel). The agreement between theory and experiment is not very good, but shows the correct trend.

The mean square displacements $\langle u^2 \rangle$ perpendicular to the surface are usually found to be about twice the bulk values [11]–[13], [17]–[19] and are larger than predicted by theory. Since bulk force constants are assumed for the surface atoms in the calculation this discrepancy suggests that force constants for surface atoms are smaller than those of bulk atoms [16].

It has to be concluded that $\langle u^2 \rangle$ and therefore also θ_D vary continuously in the surface region from one atomic layer to the next until the bulk values are reached. This problem has so far been tackled only experimentally using the fact that the penetration depth of the electrons varies with their energy. For example Jones et al. [18] performed measurements with a Ag(111) surface for a series of Bragg peaks similar to those underlying the data of fig. 10.1 for Ni(100)*). Their results also demonstrate clearly that θ_D increases with increasing energy, i.e. increasing depth of investigation. A separately determined relation between electron energy and depth of penetration was used by Jones et al. [18] in order to evaluate the variation of θ_D with the layer thickness. They found that θ increases proportionally to $(1-e^{-n})$ where n is the number of the atomic layer. However such a simple result appears somewhat surprising, since the scattering mechanism is more complicated due to multiple scattering and other effects.

Assuming certain reflection and transmission coefficients at the individual atomic layers Anderson and Kasemo [23] tried to interprete the data they obtained with a Ni(100) surface. Measurements by Reid [24] with Cu and by Tabor et al. [25] with Cr(100) and Mo(100)

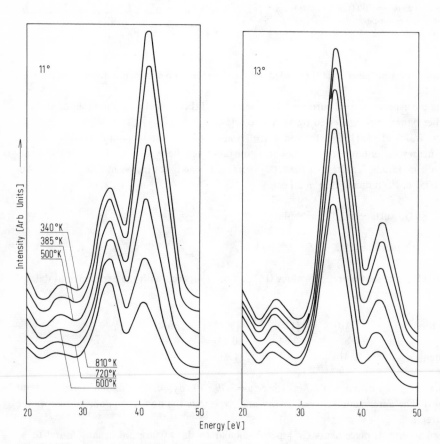

Fig. 10.3. The effect of temperature on the I/U-curve for the (0,0)-LEED beam from Cu(100). Angle of incidence: a) 11°, b) 13°. After Reid [28].

---

*) It should also be pointed out that the positions of the Bragg peaks may be shifted somewhat on the energy scale due to thermal expansions [20]. Measurements of the Debye-Waller factor are correctly done by adjusting for this energy shift [21] as was done by Jones et al. [18]. If simply ln I vs. T is plotted, it is even possible for the slope to change its sign, as shown by Lagally with Ag(111) [22].

revealed that the general trend of an increase of θ_D with increasing energy is present, but that the experimentally derived values fluctuate about a smooth curve.

The reason is to be sought in multiple scattering events. It has been shown by Somorjai and Farrell [26] that higher order diffraction processes may lead to effective Debye temperatures which are either larger or smaller than those predicted by the kinematic theory. If the mean square displacement between the first and the second atomic layer is larger than that between two layers within the bulk this effect causes a weakening of the scattering of electrons at the topmost layer and thereby an increase in the depth of penetration. Even this effect should cause a variation of the effective Debye temperature with changing electron energy [27]. The aspects of temperature effects on LEED intensities from the point of view of dynamical scattering theories have been discussed extensively by Duke and his coworkers [30].

The results of Reid [28] with Cu(100) demonstrate clearly that a kinematical analysis is not sufficient for a satisfactory explanation of the temperature effects in LEED. Fig. 10.3 shows I/U curves for the 0,0-beam at different temperatures and angles of incidence. When the two maxima between 30 and 50 eV are treated separately, plots of $\ln I$ vs. T yield values for $\langle u^2 \rangle$ and θ_D supporting a "global kinematic" interpretation. A more detailed analysis however [29] revealed that the small maximum is due to single scattering (corresponding to the 004-Bragg peak); the Debye temperature for this energy was evaluated to be 311 K which agrees reasonably well with the bulk value of Cu(330 K). The larger peak is caused by multiple scattering events, and each individual scattering process is influenced by the temperature effects. Thus the Debye temperature for this peak appears to be ~ 200 K, i.e. much lower. It is therefore clear that only a careful analysis of the scattering mechanisms involved in the formation of a particular Bragg peak enables the derived Debye temperature to be attributed to vibrations of the surface layer. Such an attempt was made by Quinto et al. [31] who studied the temperature dependence of diffracted intensity from an Al(100) surface. Fig. 10.4 shows the log I/T-plots for five Bragg maxima and the variation of the effective Debye temperature with the electron energy. As is frequently observed, θ_{eff} is about 1/2 of the bulk Debye temperature at low energy and increases with U. The most remarkable result however was that this increase was found to be strictly linear. This can be interpreted by the fact that the scattering mechanism for Al can be adequately described by a simple model for the back-scattering process [32]. If only single – backscattering events compose the essential part of a Bragg peak, the effective Debye temperature should increase monotonically with increasing energy. On the other hand, the more complicated dependence between θ_D and U in the case of Cu(100)[28] which is identified as being caused by higher orders of multiple backscattering requires calculations to higher orders. Thus a careful analysis of the temperature dependence of the diffracted intensity may enable those peaks which involve only a single reversal in the scattering path to be identified. This information can then be used for theoretical calculations of LEED intensities. The best suited computational method for this purpose appears to be Pendry's "Renormalised Forward Scattering Scheme" [33], where all forward scattering is treated exactly and the back-scattering by pertubation techniques.

The present state of the art can be summarized as follows: For all materials investigated so far the surface region is "softer" than the bulk, with a lower Debye temperature and a larger mean square displacement as predicted by relatively simple theoretical models. Details of the variation of the Debye temperature with the distance from the surface may be complicated by complex scattering processes, but these effects may be identified from a proper analysis of the variations of LEED intensities with temperature.

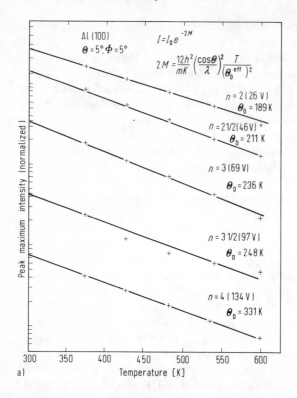

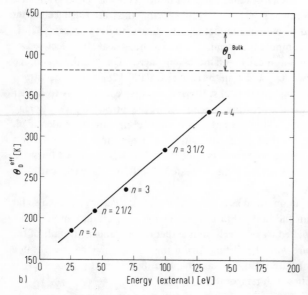

Fig. 10.4. a) The logarithm of intensity of five prominent Bragg peaks for the (0, 0)-LEED beam from Al(100) as a function of temperature. After Quinto et al. [31].

b) The effective Debye temperature as derived from fig. 10.4. a) as a function of electron energy.

10.2.2. Shifts of the Bragg maxima

Besides the general decrease of the LEED intensities increasing the temperature causes small shifts of the energies of the Bragg maxima by a few meV per degree. This effect is caused by the thermal expansion of the lattice.

For normal incidence of the primary electrons the (primary) Bragg peaks U_B (including the inner potential correction) in the I/U plot of the 0,0-beam are given by

$$2c_z = h_3 \lambda = h_3 \sqrt{\frac{150.4}{U_B}} \tag{10.8}$$

where c_z is the length of periodicity normal to the surface and h_3 an integer.

The thermal expansion coefficient β in the direction of the surface normal is defined as

$$\beta = \frac{1}{c_z} \cdot \frac{dc_z}{dT} \tag{10.9}$$

If this relation is combined with equ. (10.8) one obtains

$$\beta = -\frac{1}{2U_B} \cdot \frac{dU_B}{dT} \tag{10.10}$$

Gelatt et al. [20] and Woodruff and Seah [34] have performed detailed measurements with Ag, Ni, and Cu. The exact determination of β from peak shifts may be complicated by other effects similar to those as discussed in the preceding section. Fig. 10.5 shows some results of Wilson and Bastow [35] who measured the shifts of some Bragg peaks of the 0,0-beam from Cr(100). The resulting values vary with the electron energy, i.e. depth of information. At lower energies β reaches values of about twice the bulk value, which it approaches at higher energies. Similar results have been found for Mo [35], Ag [20], Ni [20], and Cu [34].

These findings support the idea derived from the data of surface Debye temperatures that the surface region of a solid is "softer" than the bulk. One can also try to correlate both effects [35]. The Grüneisen relation (which can be explained by means of a central force model) states that for cubic crystals the expansion coefficient is proportional to the specific heat. If it is assumed that at high temperatures the frequency of vibration is equal to the Debye frequency ω_D and $\hbar\omega_D = k_B\theta_D$ then one obtains [35] that

$$\beta \propto 1/\theta_D^2 \tag{10.11}$$

That means that

$$\beta_{surface} \cdot \theta_{D, surface}^2 = \beta_{bulk} \cdot \theta_{bulk}^2 \tag{10.12}$$

This relation has been found to be in good agreement with the experimental results for Cr and Mo [35]. The general conclusion is that the interatomic potential energy curve is shallower and more anharmonic at the surface than in the bulk, in agreement with the high surface diffusion coefficients observed and with the concept of weaker binding at the surface.

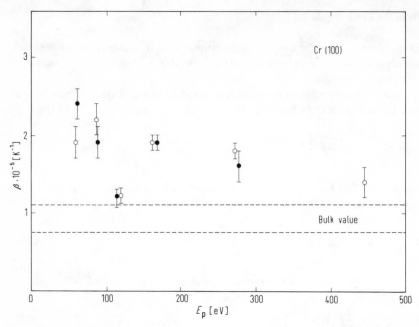

Fig. 10.5. Variation of the thermal expansion coefficient β (as derived from shifts of the Bragg maxima) for Cr(100) with the electron energy. After Wilson and Bastow [35].

10.2.3. Thermal diffuse scattering

If the temperature of a surface is raised the intensities of the LEED diffraction spots decrease, but simultaneously the brightness of the background increases, which means that the total elastically backscattered intensity is not noticeably altered, but only the directions of the electrons. At higher temperatures electrons are deflected from the Bragg directions by interactions with the vibrating lattice and contribute to the intensity of the diffuse background.

In order to derive a more quantitative understanding the starting point is again equ. (10.1) describing the time dependent position $r_j(t)$ of an atom j:

$$r_j(t) = {}^{\circ}r_j + u_j(t) \tag{10.1}$$

Here it must be realized that the displacements $u_j(t)$ are not any functions of time, but are of the form

$$u_j(t) = u_0 \exp\left[i(p \cdot r_j - \omega_p t)\right] \tag{10.13}$$

This equation expresses the fact that a wave of elastic deformation propagating in the crystal leads to periodic displacements of the individual lattice atoms in space and in time, as in the case of sound propagation. This propagation of the displacement of the lattice atoms can also be treated as the movement of a phonon with energy $\hbar\omega_p$ and momentum p.

Electrons interact inelastically with the lattice and thereby either create or destroy phonons (i.e. excite or extinguish lattice waves). In both processes the energy E and momentum k of the electrons change according to $k' = k \mp p$ and $E' = E \mp \hbar \omega_p$. The variation of the electron's momentum is equivalent to a change of its direction, which means that these electrons do not contribute to the intensity of the elastically scattered Bragg beams.

The probability of interaction between electrons penetrating a crystal surface and phonons increases with increasing temperature, since the density of phonons increases with the temperature according to the Boltzmann factor. If it were possible to separate the strictly elastically back-scattered electrons from all other electrons, then this current would decay exponentially with temperature.

Since the energy losses and energy gains by phonon interactions are only of the order of a few tens of meV these "quasielastic" electrons are registered together with the true elastic electrons in normal LEED equipment because of instrumental resolution. These electrons can be divided into three parts:

a) The zero-phonon portion corresponding to the electrons which have been scattered purely elastically.

b) The one-phonon part, arising from electrons which have been scattered once by a phonon.

c) The multi-phonon portion, comprising all electrons which have interacted more than once with phonons.

Detailed experimental investigations on thermal diffuse scattering have been performed by Webb and his coworkers [37]–[40]. Fig. 10.6 shows the intensity of "elastically" backscattered electrons from Ni(111) as registered by the total LEED screen current as a function of the

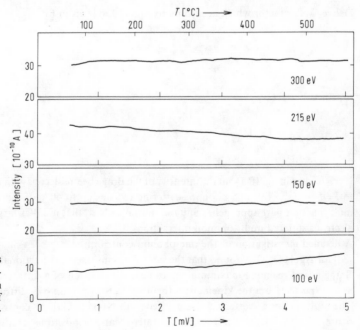

Fig. 10.6. Total intensities of electrons back-scattered "elastically" from Ni(111) as a function of temperature. After Lagally and Webb [37].

temperature. This total intensity does not depend on the temperature, although the intensity of the Bragg beams decreases by a factor of 30 within the same temperature range [37]. The distribution of the electrons in the three types listed above is shown in fig. 10.7. $2W$ is the

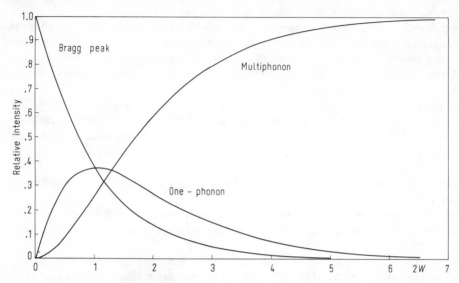

Fig. 10.7. Distribution of back-scattered "quasielastic" electrons into zero-phonon, one-phonon and multiphonon parts, depending on the Debye-Waller factor $2W$. After Lagally and Webb[37].

Debye-Waller factor which according to equ. (10.6) is given by

$$2W \propto \frac{1}{\lambda^2} \cdot \frac{T}{\theta_D^2} \tag{10.13}$$

or, since $\lambda^2 \propto 1/U$

$$2W \propto \frac{U \cdot T}{\theta_D^2} \tag{10.14}$$

It can be seen in fig. 10.7 that the intensity of the Bragg beams decays with temperature according to $I = I_0 \cdot e^{-2W}$. The one-phonon scattering reaches up to 40% of the total intensity and the multi-phonon part approaches unity monotonically so that if $2W$ is large enough all "elastic" electrons suffered multi-phonon interactions.

A detailed investigation of the one-phonon contribution has been made by McKinney et al. [38] for Ag(111). They found that the intensity is inversely proportional to the distance from a Bragg beam, reaching maximum values near the position of a diffraction spot. This result is to be expected from the kinematical theory as is the case in x-ray diffraction [9].

The multi-phonon scattering was studied with Ni(111) [39]. In contrast to the one-phonon scattering this intensity is not concentrated near the positions of the Bragg beams but is uniformly distributed throughout space. It is thus evident, why the diffraction spots enlarge and why the background intensity increases with increasing temperature.

A series of further observations as usually made in LEED experiments can be understood from equ. (10.3). If the accelerating voltage is increased the background becomes brighter, although the temperature remains constant. This is caused by the Debye-Waller factor being proportional to U as well as to T. Some materials exhibit a priori a high background intensity due to low values of the Debye temperature θ_D.

In the case of nickel at room temperature for $U = 100$ eV the Debye-Waller factor is about 1.2. According to fig. 10.7 this corresponds to roughly equal amounts of the three contributions [38]. If LEED intensities are measured by the usual Faraday-cup arrangement the intensity of a Bragg-beam contains about 25% of thermal diffuse scattered electrons, and in the case of Ag under the same conditions up to about 40% [40]. Studies of thermal diffuse scattering of slow electrons do not yield detailed information about particular surface processes, but the knowledge is essential for an understanding of LEED phenomena. An extensive theoretical treatment with the inclusion of dynamical scattering effects has been given by Duke [41].

10.3. High resolution energy loss spectroscopy

The excitation energies of surface vibrations lie well below 0.5 eV and it is therefore impossible to detect the corresponding energy losses with normal LEED display systems or with the electron spectrometers customarily used for Auger electron spectroscopy. The necessary energy resolution of about 10 meV can be achieved using electrostatic cylinder condensors, e. g. of the type described by Ehrhardt et al. [42] to monochromate the primary electrons as well as analysing the secondary electrons. The impact energy used should be only a few eV which imposed additional difficulties on the construction and handling of such spectrometers. This technique was first used with success by Propst and Piper [8] who were able to record spectra of the vibrational states of particles adsorbed on W(100) surfaces.

More advanced spectrometers have been constructed by Ibach [5], who discovered the existence of surface phonons at ZnO [5]. Fig. 10.8 shows recorder traces of energy-loss spectra from a ($1\bar{1}00$) ZnO surface at two different temperatures using a primary energy of 7.5 eV. To the right of the elastic peaks (i. e. energy loss = 0) four equally-spaced peaks are clearly discernible which must be attributed to the creation of 1, 2, 3, and 4 phonons respectively. In the spectrum which was taken at the higher sample temperature an additional peak appears which is due to the energy gain of the electron by destroying a phonon. The intensities of energy-gain and energy-loss peaks are correlated through the Boltzmann factor $\exp(-\hbar\omega_p/kT)$, $\hbar\omega_p$ being the energy of the phonon. These phonons were interpreted by a continuum theory for surface phonons in ionic crystals as derived by Fuchs and Kliewer [43] whereby their energy depends simply on the dielectric properties. The theoretical value was calculated to be 69 meV which is in excellent agreement with the experimental result of 68.8 ± 0.5 meV. According to the theory of Lucas and Sunjic for this "dipole scattering" [6], the intensity of the one-phonon loss should be proportional to $E_0^{-1/2}$, where E_0 is the primary energy of the electrons, which was also found to be in very good agreement with experiment [4]. A more refined theory has recently been developed by Evans and Mills [36].

As outlined in the foregoing section the one-phonon loss should be located close to the directions of the elastically reflected beams. This has also been confirmed experimentally.

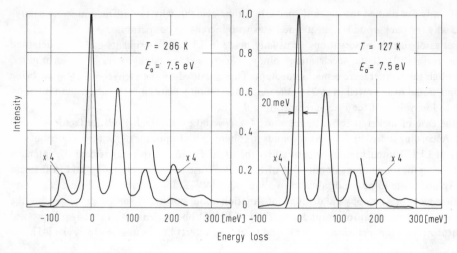

Fig. 10.8. High resolution energy loss spectra for electrons specularly reflected from a $(1\bar{1}00)$ ZnO surface. After Ibach [5].

Similar experiments have been performed by Ibach [44] with a Si(111) surface where an energy loss at 56 meV was found. Although the detailed nature of this phonon is not clear it is interesting to note that this type of interaction of electrons with matter which is considered to be dipole scattering is possible even from surfaces of non infrared active materials.

The mechanism of this type of electron/solid interaction can be explained by a scattering process apparently produced by the electric field outside the crystal generated by atomic motions. The vibration of a surface atom perpendicular to the surface may produce an oscillating

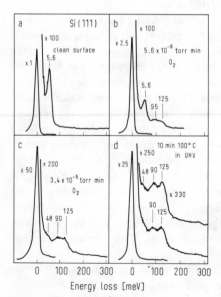

Fig. 10.9. Energy loss spectra of 5 eV electrons reflected from a (111) Si surface at different stages of oxidation. After Ibach [4a].

electric dipole moment. This model would seem to have immediate application to the vibrations of adsorbed particles. The energies of vibrations in a two-atomic molecule are typically in the range of a few 100 meV (=wavelength of a few microns).

After the preliminary measurements by Propst and Piper [8] Ibach [4a] performed similar experiments with a Si(111) surface after exposure to oxygen. The variations of the spectra under the influence of oxygen adsorption are shown in fig. 10.9 demonstrating the appearance of three new losses at 48, 90 and 125 meV. Although a unique model for the state of the adsorbed oxygen cannot be made in this case, these results offer an exciting prospect for the study of adsorbed particles. Further experiments of this type have been performed by Adnot et al. [45] with adsorbate covered stainless steel-samples.

Ibach [4] estimated that a total number of 10^{11} to 10^{12} scattering particles should be enough for acceptable measurements which represents a sensitivity orders of magnitude better than for infrared spectroscopy. Furthermore the theoretical treatment of dipole scattering would appear to be comparatively simple since the intensity always refers to the elastically scattered intensity without any complications due to inelastic damping or multiple scattering [4]. On the other hand this technique still presents considerable instrumental problems so that it will probably remain restricted to a few laboratories in the near future, in spite of its enormous promise.

10.4. Infrared spectroscopy

An alternative possibility to excite vibrations at surface is to use infrared radiation. Since electromagnetic waves penetrate deeply into solids the surface sensitivity is a priori rather restricted. IR spectroscopy however has been applied very successfully to the study of adsorbed species on materials with a high surface to volume ratio such as finely dispersed supported catalysts. This topic has been treated extensively by Little [46] and Hair [47]. The aim of such investigations is to identify the chemical nature of the adsorbed species, to obtain information of variations in the chemical bond within the molecules from characteristic frequency shifts, and finally to elucidate the nature of the bond between surface and adsorbate.

With clean surfaces, mainly metallic films evaporated in UHV, only the vibrational bond C=O of adsorbed carbon monoxide has been investigated with some success. Bradshaw and Pritchard [48] found two adsorption bands at 2050 and 1900 cm^{-1} after CO adsorption at a thin nickel film ($d < 200$ Å), a result which was confirmed in later investigations [49]. A disadvantage of this transmission technique is that only very thin films can be used since they must be transparent to IR radiation. The transmitted intensity varies by only a few percent after adsorption.

Better experimental conditions and higher sensitivity may be achieved by reflectance spectroscopy as proposed by Greenler [50].

The optimum number of reflections and the optimum angle between beam and surface depends on the optical constants of the material [51]. Even a single reflection at one surface may provide sufficient sensitivity.

Using this technique Yates and King [52] were able to detect CO adsorbed on polycrystalline tungsten below monolayer coverages. Comparison of the data with the results of flash desorption experiments revealed that only the more weakly bound α-state of adsorbed CO was registered, the band shifting with increasing coverage towards higher frequencies.

Similar experiments have been performed by Kottke et al. [53] with CO adsorbed on evaporated Au surfaces. The bands shifted to lower frequencies with increasing coverage and as little as 5% of a monolayer can be detected in their spectra. The first measurements with single crystal surfaces were reported by Pritchard [54] for the system CO/Cu. The results were correlated with LEED and surface potential measurements.

Infrared spectroscopy shows some promise for the study of adsorption under well-defined conditions, but it remains to be seen whether enough sensitivity can be developed to allow measurements on anything other than the very strongest chromophores.

10.5. References

[1] T. Wolfram, R. E. De Wames and E. A. Kraut, J. Vac. Sci. Techn. 9, 685 (1972).
[2] a) T. S. Chen, R. E. Allen, G. P. Allredge and F. W. de Wette, Solid State Comm. 8, 2105 (1970);
    Phys. Rev. Lett. 26, 1543 (1971).
    b) R. E. Allen, G. P. Allredge and F. W. de Wette, Phys. Rev. Lett. 24, 301 (1970).
[3] A. A. Maradudin and F. Wallis, Phys. Rev. 148, 962 (1966).
[4] H. Ibach,
    a) J. Vac. Sci. Techn. 9, 713 (1972);
    b) Festkörperprobleme (Ed. O. Madelung) 11, 135 (1971).
[5] H. Ibach, Phys. Rev. Lett. 24, 1416 (1970).
[6] A. A. Lucas and M. Sunjic, Phys. Rev. Lett. 26, 229 (1971).
[7] a) L. Dobrzynski, Surface Sci. 20, 99 (1970).
    b) G. Armand, P. Masri and L. Dobrzynski, J. Vac. Sci. Techn. 9, 705 (1972).
[8] F. M. Propst and T. C. Piper, J. Vac. Sci. Techn. 4, 53 (1967).
[9] See for example: R. W. James: The optical principles of the diffraction of X-rays. Bell & Sons, London 1962.
[10] A. U. McRae and L. H. Germer, Phys. Rev. Lett. 8, 489 (1962).
[11] K. Christmann, O. Schober and G. Ertl, unpublished.
[12] A. U. McRae, Surface Sci. 2, 522 (1964).
[13] J. M. Morabito, R. F. Steiger and G. A. Somorjai, Phys. Rev. 179, 638 (1969).
[14] A. A. Maradudin, Solid State Phys. 19, 1 (1966).
[15] R. F. Wallis, B. C. Clark and R. Herman in: The Physics and Chemistry of Solid Surfaces. (Ed. G. A. Somorjai) Wiley, New York 1969, p. 17-1.
[16] R. F. Wallis, B. C. Clark and R. Herman, Phys. Rev. 167, 625 (1968).
[17] P. Masri and L. Dobrzynski, Surface Sci. 32, 623 (1972).
[18] E. R. Jones, J. T. McKinney and M. B. Webb, Phys. Rev. 151, 476 (1966).
[19] R. M. Goodman and G. A. Somorjai, J. Chem. Phys. 52, 6325 (1970).
[20] C. D. Gelatt, M. G. Lagally and M. B. Webb, Bull. Am. Phys. Soc. 14, 793 (1969).
[21] D. P. Woodruff and M. P. Seah, Phys. Lett. 30 A, 263 (1969).
[22] M. G. Lagally. Z. Naturforsch. 25 a, 1567 (1970).
[23] S. Anderson and B. Kasemo, Solid State Comm. 8, 1885 (1970).
[24] R. J. Reid, Surface Sci. 29, 623 (1972).
[25] D. Tabor, J. M. Wilson and T. J. Bastow, Surface Sci. 26, 471 (1971).
[26] G. A. Somorjai and H. H. Farrell, Adv. Chem. Phys. 20, 215 (1971).
[27] G. E. Laramore and C. B. Duke, Phys. Rev. B2, 4783, 4765 (1970).
[28] R. J. Reid, Phys. Stat. Solidi (a) 4, Ks11 (1971).
[29] B. W. Holland, Surface Sci. 28, 258 (1971).
[30] a) C. B. Duke, N. O. Lipari and U. Landman, Phys. Rev. B8, 430 (1973).
    b) C. B. Duke, Adv. Chem. Phys. 27, 1 (1974).
[31] D. T. Quinto, B. W. Holland and W. D. Robertson, Surface Sci. 32, 139 (1972).
[32] S. Y. Tong and T. N. Rhodin, Phys. Rev. Lett. 26, 711 (1971).
[33] J. B. Pendry, Phys. Rev. Lett. 27, 856 (1971).

[34] D.P. Woodruff and M.P. Seah, Phys. Stat. Solidi (a) *1*, 429 (1970).

[35] J.M. Wilson and T.J. Bastow, Surface Sci. *26*, 461 (1971).

[36] E. Evans and D.L. Mills, Phys. Rev. *B5*, 4126 (1972); *B6*, 3163 (1972).

[37] M.G. Lagally and M.B. Webb, in: The structure and chemistry of solid surfaces (Ed. G.A. Somorjai) Wiley, New York 1969, p. 20-1.

[38] J.T. McKinney, E.R. Jones and M.B. Webb, Phys. Rev. *160*, 523 (1967).

[39] R.F. Barnes, M.G. Lagally and M.B. Webb, Phys. Rev. *171*, 627 (1968).

[40] M.G. Lagally, Proc. Int. LEED Summer School, Smolenice (CSR), Prague 1972, Vol. 1, p. 166.

[41] C.B. Duke, Proc. Int. LEED Summer School, Smolenice (CSR), Prague 1972, Vol. 2, p. 7.

[42] H. Ehrhardt, L. Langhans, F. Linder and H.S. Taylor, Phys. Rev. *173*, 222 (1968).

[43] R. Fuchs and K.L. Kliewer, Phys. Rev. *140*, A2076 (1965); *144*, 495 (1966).

[44] H. Ibach, Phys. Rev. Lett. *27*, 253 (1971).

[45] A. Adnot, Y. Ballu and J.D. Carette, J. Appl. Phys. *43*, 2796 (1972).

[46] L.H. Little: Infrared spectra of adsorbed species. Academic Press, New York 1966.

[47] M.L. Hair: Infrared Spectroscopy in Surface Chemistry. Marcel Dekker, New York 1967.

[48] A.M. Bradshaw and J. Pritchard, Surface Sci. *17*, 372 (1969).

[49] A.M. Bradshaw and O. Vierle, Ber. Bunsenges. *74*, 630 (1970).

[50] R.G. Greenler, J. Chem. Phys. *50*, 1963 (1969).

[51] H.G. Tompkins and R.G. Greenler, Surface Sci. *28*, 194 (1971).

[52] J.T. Yates and D.A. King, Surface Sci. *30*, 601 (1972).

[53] H.L. Kottke, R.G. Greenler and H.G. Tompkins, Surface Sci. *32*, 231 (1972).

[54] J. Pritchard, J. Vac. Sci. Techn. *9*, 895 (1972).

11. Processes in adsorbed layers

11.1. Introduction

This last chapter provides a short summary of the possible information which is to be obtained about processes at solid surfaces in contact with a gas phase using the experimental techniques described. These processes are always associated with adsorption which may, under certain circumstances, proceed farther to the formation of three dimensional compounds with the solid (e. g. in oxidation) or to the growth of multi-atomic layers (e. g. in epitaxy). In addition changes in the chemical nature of the adsorbed particles are the basis of heterogeneous catalysis.

A complete description of an adsorbed layer essentially consists of the following information:

a) The chemical composition of the surface. An elemental analysis of a surface may (with the exception of hydrogen) best be achieved by means of Auger electron spectroscopy, although x-ray photoelectron spectroscopy should also be useful for this purpose. Much more complicated is the derivation of the molecular composition of the adsorbed species. Again PES might help, but there are not yet enough data available to be sure of the capabilities of the method. In principle infrared spectroscopy of characteristic bands can distinguish functional groups. Further information may be obtained from field ion mass spectroscopy [1] or from secondary ion mass spectroscopy (SIMS) [2], although with these techniques some caution is needed, since, for example, conclusions about the chemical nature of the surface are drawn from the analysis of particles in the gas phase. Promising aspects are also offered by low energy ion backscattering, a technique which is furthermore able to yield some structural information [79].

b) The surface concentration or coverage θ*). The relative amount of adsorbed particles can frequently be determined by measuring a quantity which can confidently be assumed to be or has been shown to be proportional to the coverage. Within certain limits the AES peak-to-peak height, the change of the work function or the area below a flash-desorption peak [3] can be used for this purpose. Ellipsometry [4] needs some further experimental development and presents special problems in the submonolayer region.

Calibration of the relative data to absolute coverages can be achieved sometimes from those LEED patterns which allow the unequivocal derivation of the geometrical configurations of adsorbed particles with respect to each other [5]. The direct determination of absolute adsorbed quantities is relatively easy with large area adsorbents like evaporated films or porous solids for example by volumetric methods but can in principle also be performed on single crystal planes of small area. This may be achieved by calibrating the evaporation rate of metallic adsorbates (e. g. by means of a quartz microbalance) or with a radioactive tracer [6] in suitable cases. The roughness of the surface must be taken into consideration if the number of adsorbed particles per square centimeter is required. With large-area samples the usual techniques of surface area determination may be used.

Certainly in much single-crystal work determining the absolute coverage is still a serious problem.

---

*) The coverage θ is defined in this context as the ratio of the number of adsorbed articles to the number of surface atoms in the topmost substrate layer.

c) *The configuration of the adsorbed particles with respect to the substrate atoms ('adsorption sites') and with respect to each other.* These problems have to be solved by means of LEED. However so far this is possible only to a limited extent, mainly because of the difficulties involved in the analysis of intensity data.

d) *The binding energy of the adsorbed particles to the surface.* Integral heats of adsorption may be derived approximately from the activation energies for desorption, which can be evaluated from an analysis of flash desorption spectra [3]. Calorimetric measurements of the adsorption energies can be made when using thin evaporated films [7]. In the case of single crystal planes the isosteric heats of adsorption are frequently used, being obtained via the Clausius-Clapeyron equation

$$\frac{\mathrm{d}\ln p}{\mathrm{d}(1/T)}\bigg|_{\theta=\mathrm{const}} = -\frac{E_{ad}}{R} \tag{11.1}$$

provided that a reversible adsorption-desorption equilibrium is established. In this case it is necessary to have a parameter for the coverage θ which is independent of temperature. Measurements of this type have for example been made using the transformation of a LEED pattern as coverage monitor [8]. Of more general application is the work function change on adsorption which can be measured with considerable accuracy. In this way the differential isosteric energy of adsorption as a function of coverage may be determined.

e) *The interaction energies between adsorbed particles.* The type of interaction (attractive or repulsive) determines the configuration of the adsorbed particles with respect to each other and is therefore reflected in the LEED patterns. Estimates of the magnitudes of these energies can be derived from a knowledge of $E_{ad}(\theta)$ and from observations of order/disorder transitions.

f) *The charge distribution in the adsorbate complex.* Measurements of work function changes may yield the sign of charge transfer between the substrate and the adsorbed particles. Derivations of the dipole moment of the adsorbate or even of the charge distribution are however much more complicated.

g) *The energies and energy distributions of chemisorption induced orbitals.* This question is of fundamental importance in any attempt to understand the nature of the adsorption bond and is a major field of application of photoelectron spectroscopy. Also ion neutralization spectroscopy, field emission spectroscopy and energy loss spectroscopy are very useful in this context.

h) *Vibrational states of the adsorbate complex.* The frequencies of the surface vibrations may be obtained by means of high-resolution energy loss spectroscopy (analysis of the "quasielastic" electrons) and by infrared spectroscopy.

i) *Kinetics of adsorption and desorption.* Any technique which allows recording of the surface concentration or of the gas phase composition (e.g. a mass spectrometer) is useful for this purpose.

k) *Surface mobility.* Surface diffusion processes may be followed by field emission microscopy and to some extend also by LEED (spot sharpening) or scanning AES. Very refined and detailed information on residence times, rotation frequencies etc. follow from the application of nuclear spin relaxation techniques, which unfortunately has been restricted to weakly adsorbed molecules on large-area nonmetallic materials [9] in the past for experimental reasons.

l) Mechanisms and kinetics of chemical reactions in the adsorbed layer. Desorbing reaction products can be analyzed by a mass spectrometer. Conclusions on processes in the adsorbed state can be drawn from combined observations with different techniques.

Certainly this list could be continued to include other important points, in particular variations of solid state properties or energy and momentum transfer between the surface and colliding particles from the gas phase, as studied for example by molecular beam techniques. Even for any individual adsorption system we are far from being able to present a complete solution of the above mentioned problems. It is evident that this task may only be tackled by combined application of several experimental methods, whereby those described in this work are the most important. A very serious consideration in all of these studies is the possible effect which low energy electrons themselves may have on the system under investigation, since the very property which makes them ideal for surface work, their strong interaction with matter, may conceivably result in undesired changes to any adsorbate present.

In the following sections some different aspects of processes in adsorbed layers are treated by illustrating the results obtained in some examples selected more or less subjectively from the wealth of information in the literature.

11.2. Reactions of electrons with adsorbed particles

Several of the methods for surface studies are based on the fact that impinging electrons excite or eject bound electrons. It is therefore a quite natural consequence that even chemical bonds, either those within adsorbed molecules or those between the surface and adsorbed particles, may be influenced by incident electrons. In the ion source of a mass spectrometer this effect causes the well known fragmentation of molecules.

In the case of adsorbed layers this may be very disturbing, since with all methods based on electron impact changes in the subject of the investigation must be considered. And indeed in a series of studies degradations or transformations of LEED patterns or chemical changes such as the formation of carbon from adsorbed CO have been observed [10]. Particular care must be taken if adsorbed layers are investigated with AES using the retarding field technique, since the current densities are rather high. Thermal effects caused by the electron beam current may also be important in these cases. However in most LEED studies and in AES work with low primary currents (using the cylindrical mirror analyzer) the effects of the electron beam on the nature and concentration of adsorbed particles may be neglected, indicating small values for the reaction cross section of the corresponding processes. General experience indicates that the influence of the electron beam becomes more severe when the adsorbed particles are only weakly bound to the surface. Metallic adsorbates are almost unaffected. Although the effect of electrons on adsorbed layers is normally regarded as a nuisance, the desorption processes stimulated by electron impact may also be followed in detail leading to a kind of spectroscopy for the characterization of adsorbates.

It has been recognized for a long time that positive ions as well as neutral particles may be desorbed from adsorbate covered surfaces under electron impact [11]. Systematic investigations were performed for the first time in 1964 by Menzel and Gomer [12] and by Redhead [13]. In principle two types of measurements may be performed. In the first of these a surface parameter is measured which depends on the nature and the concentration of adsorbed particles. This method was used by Menzel and Gomer [12] who studied the variation of the work function in an FEM apparatus using the system CO/W. The other possibility consists in the detection of the particles released from the surface, as was done by Redhead [13]. He measured the total ion emission (without mass selection) as well as the kinetic energies of the ions using a retarding field technique. In further developments the detection of desorbed particles was performed by mass spectrometry as originally done by Moore [14]. Menzel [15] studied the system CO/W in an apparatus as shown in fig. 11.1 which allowed the measurement of ions as well as of neutrals.

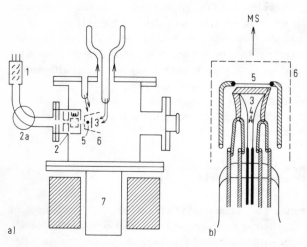

Fig. 11.1. a) Scheme of an apparatus for the study of electron stimulated desorption of ions and neutrals.
b) View of the anode section: 1: electron multiplier, 2: ion source, and 2a: magnet of the mass spectrometer, 3: anode (W foil), 4: potential wires, 5: cathode filament, 6: shielding grid, 7: pump.
After Menzel [11a].

If the primary current is small enough so that the number of adsorbed particles is not appreciably altered during the experiment, then the number of desorbing particles per second is found to be proportional to the electron current and to the coverage of adsorbed particles in that particular state. This technique can therefore be used to monitor the surface concentration e.g. in experiments where the adsorption kinetics are studied, as for example done by Lichtmann et al. [16] with the Ni(100)/H_2 system.

If the probability for desorption w is defined as the ratio of the flux of desorbing particles to the flux of exciting electrons, $w = i_{des}/i_e$, then w is found to be in the range of 10^{-4} to 10^{-8} for ions and $\leqslant 10^{-2}$ for neutrals for primary electron energies of about 100 eV (which exhibit maximum yields) [15]. If the primary electron current is much higher then the concentration of adsorbed particles can be observed to be depleted by a first-order reaction. From such studies the cross sections q, being defined as $q = w/n_s$, for the individual processes can be determined. It has been found that different adsorbed states may differ considerably in their cross sections. The maximum values of q are in the range of 10^{-18} to 10^{-21} cm^2 and are therefore orders of magnitudes smaller than those for the corresponding processes with free molecules, which are usually in the range of 10^{-15} cm^2.

Besides having different cross sections different adsorption states are further distinguished by their desorption products [15] or by the varying kinetic energies of the desorbing particles [17]. As can be seen from fig. 11.2 the desorption starts at particular threshold energies and increases with increasing electron energy, until at about 100 eV maximum q values are obtained. In this example [15] from the $\alpha-CO$ state on polycrystalline tungsten CO^+- as well O^+-ions are ejected with different threshold energies, pointing to differences in the corresponding primary processes. Electron impact may furthermore also cause a conversion from one adsorbed state to another, as for example observed for the $\alpha-CO$ state on $W(100)$ [18].

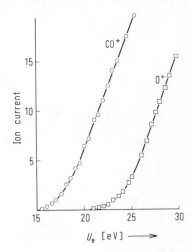

Fig. 11.2. Ion currents of CO^+ and O^+ desorbing from a CO covered tungsten surface under electron impact as a function of the electron energy. After Menzel [11a].

In order to explain the observed phenomena Menzel and Gomer [12] and Redhead [13] proposed independently the following mechanism: Electron impact causes Franck-Condon transitions of bound electrons in the adsorbate complexes into excited states from which there is a probability of dissociation of the surface complex with subsequent total or partial desorption of the particle. The existence of such electronically excited surface complexes is also suggested by the results of electron loss spectroscopy.

The decomposition of the excited surface complex competes with the filling of the empty ground state level via electron tunneling from the metal substrate. Obviously this process is very effective since the chemisorption orbital strongly couples with the metallic valence band and therefore (fortunately!) the cross section q is much lower than for free molecules. This mechanism also provides a qualitative explanation for the larger q values found for weakly bound particles. More detailed studies [15] revealed that the desorption of neutral particles does not proceed via the neutralisation of ions [13] but through the formation of an anti-bonding state. It is to be expected that neutral particles are also released from the surface in electronically excited states. This has been nicely demonstrated by Newsham and Sandstrom [19] who used a time of flight apparatus [20] which allowed the detection of excited neutrals as well as measurements of their lifetimes.

It is also of interest that the cross section for photodesorption has been found to be smaller than in electron impact desorption ($5 \cdot 10^{-21}$ cm^2 at $h\nu=5$eV) for the CO/W system. This effect is probably due to a similar electron transition from the metal to the empty ground state level during the lifetime of the excited state [21].

11.3. Ordered adsorbed phases

It has become clear over the years of LEED studies that the formation of ordered adsorbed phases is more the rule than the exception. In order to understand this phenomenon it is necessary to consider the energetic conditions for adsorbed particles on a single crystal surface in more detail.

If a particle approaches a surface from the gas phase it will experience a potential of the type shown in fig. 11.3a. (When adsorption is associated with dissociation there may be an

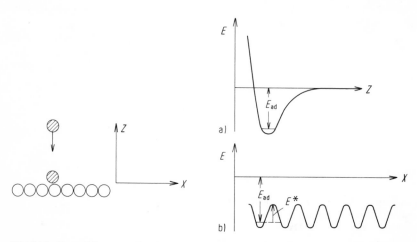

Fig. 11.3. Potential energy diagram for a particle adsorbed on a single crystal surface. a) Variation of the energy with the distance z normal to the surface. b) Variation in the direction x parallel to the surface for optimum values of z.

activation energy barrier to surmount as well). The depth of the potential trough depends on the location with respect to the surface atoms and consequently varies periodically across the surface (fig. 11.3b). The places with minimum energy are then the "adsorption sites". The adsorbed particles vibrate in all directions and will desorb if they accumulate enough thermal energy from the lattice for the energy barrier in the direction normal to the surface to be overcome. The residence time of the particle on the surface is given by the so called Frenkel equation [22]

$$\tau = \tau_0 \cdot e^{E_{ad}/kT} \tag{11.2}$$

where E_{ad} is the adsorption energy and τ_0 is about as long as a single vibration, i.e. 10^{-13} s, a value which may be modified by entropy effects.

The potential barriers shown in fig. 11.3b indicate the activation energy E^* for surface diffusion of the adsorbed particle in a chosen direction (parallel to the surface). There will always be specific directions of the surface with minimum activation energy E_0^*, depending on the particular nature of the adsorption bond. In all systems so far investigated E_0^* is always smaller than $\sim 0.5\, E_{ad}$ and may have even much lower values. The 'lifetime' on a specific adsorption site

$$\tau' = \tau_0' e^{E_0^*/kT} \tag{11.3}$$

is therefore always very much smaller than the total residence time τ on the surface. As a consequence the adsorbed particle changes its position on the surface by surface diffusion many times prior to desorption provided that τ is not merely of the duration of a few vibrations and the adsorbed particle is thermally accommodated with the surface.

If $E_0^* \gtrsim 10 \, kT$ the adsorbed particle will be nearly immobile on the particular adsorption site; only if the thermal energy kT exceeds a value of about $0.1 \, E_0^*$ will the particle jump from one adsorption site to a neighbouring one. However, so long as $E_0^* > kT$ the adsorbed particle will be resident on an adsorption site for most of the time. If $E_0^* < kT$ the particle will move quite freely over the surface like a two-dimensional gas, i.e. there is no longer localization to certain 'adsorption sites'.

The situation is different if not a single particle but rather a whole ensemble is adsorbed on the surface, since mutual interactions between the adsorbed particles come into play. These

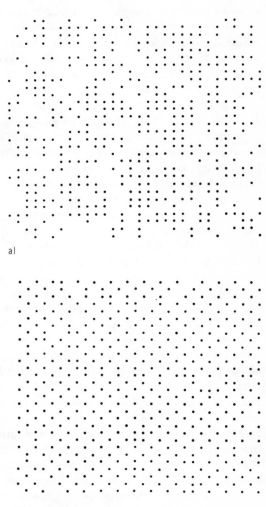

a)

b)

Fig. 11.4. Computer simulation of the geometric configuration of adsorbed particles at $\theta = \frac{1}{2}$ on a 30×30 lattice [26].
a) random distribution, b) equilibrium configuration in the case of weak repulsive interactions ($\varepsilon = 2.2 \, kT$) between nearest neighbours [26].

interactions can be either repulsive or attractive and can also operate indirectly, that is via the valence electrons of the substrate atoms [23]. If for the sake of argument $E_0^* > kT$ is assumed, then for attractive interactions the formation of adsorbate covered islands on the surface separated by bare surface areas can be expected. As the coverage is increased the islands grow together. Such cases are characterized by LEED 'extra' spots which are sharp even at low coverages and display an adsorption energy which is independent of coverage almost up to saturation of the particular adsorbed phase. If the interactions are predominantly repulsive the heat of adsorption decreases continuously with increasing coverage and at low surface concentrations the adsorbed particles tend towards a uniformly spaced distribution. Above a certain coverage the formation of ordered arrangements in registry with the substrate lattice should occur (always provided that $E^* > kT$), a situation which is referred to as a "lattice gas" [24].

The formation of the equilibrium configuration of such a "lattice gas" can be simulated in a computer experiment using the Monte Carlo method [25]. If a square lattice consisting of 30×30 adsorption sites is covered at random up to half a monolayer (corresponding to the statistical arrival of particles from the gas phase) the surface configuration will be like that shown in fig. 11.4a. If a weak repulsive interaction between neighbouring particles is now assumed and the particles are allowed to diffuse on the surface the total energy as well as the configuration entropy will continuously decrease (fig. 11.5) until the equilibrium configuration is reached. Such an equilibrium configuration is plotted in fig. 11.4b and represents clearly a $c2 \times 2$-structure containing characteristic structural defects (vacancies, interstitials, and dislocations) such as are common in three-dimensional lattices.

The degree of ordering is directly connected with the shapes and the relative intensities of the LEED 'extra' spots, and depends on the magnitude of ε/kt, where ε is the interaction energy. Order-disorder transitions are therefore to be expected as the temperature is raised. Such phenomena have been observed in a series of systems. The simplest case, a $c2 \times 2$-structure, can be treated either by the Monte Carlo-technique [26] or by adopting the exact solution of the two-dimensional Ising model [27] and is illustrated for the $H_2/W(100)$ system in fig. 11.6, which shows the experimental values of the relative LEED intensities of the $(\frac{1}{2}, \frac{1}{2})$ spot as measured by Estrup [28] together with a theoretical curve calculated in the kinematic approximation assuming weak repulsion between nearest neighbours. The critical temperature is near 500 K which corresponds to an interaction energy of about 0.1 eV or around 10% of the adsorption energy. Theoretical evaluations of the interaction energies [23] generally result in a value about an order of magnitude smaller than the adsorption energy. In most cases however order-disorder transitions in the adsorbed phase cannot be observed because of the onset of thermally induced desorption. In the case of repulsive interactions prior to the formation of long range order structures, only small ordered domains are formed giving fuzzy LEED extra spots.

If the coverage increases further the two-dimensional pressure of the lattice gas increases and even if $E^* \approx kT$ the motions of the adsorbed particles become restricted by lateral interactions until finally a closed dense structure is formed, probably with the formation of liquid-like structures as intermediates. The adsorption sites of the lattice gas are abandoned at this stage. The substrate either determines only the orientation of the surface layer to give an incoherent structure, or partial registry between the overlayer and the substrate lattice occurs as the result of a compromise between the adatom-adatom and the adatom-substrate interactions. The thermodynamic equilibrium configurations are not realized in all cases. As in the formation

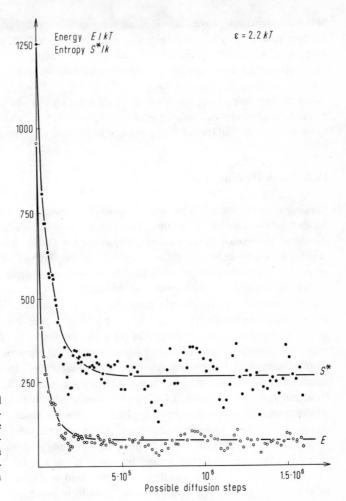

Fig. 11.5. Variation of the total energy and of the configurational entropy during surface diffusion starting with the random distribution of fig. 11.4a and assuming repulsive interactions $\varepsilon = 2.2 \ kT$ between nearest neighbours [26].

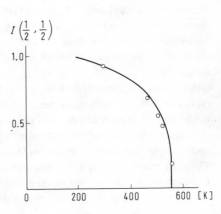

Fig. 11.6. Order-disorder transition for the $c\,2 \times 2$-structure of hydrogen adsorbed on W(100). LEED intensity of the $(\frac{1}{2}, \frac{1}{2})$ 'extra' spot as a function of temperature. Experimental data (circles) taken at 45 V. The solid curve gives the theoretical values for the two-dimensional Ising model with a critical temperature $T_c = 550$ K. After Estrup [28].

of three-dimensional phases, kinetic phenomena such as low diffusion coefficients or nucleation phenomena may play an important role and lead to metastable states. (Such a situation can even occur with clean surfaces, e. g. in the case of Ge and Si(111) where the metastable structures of the cleaved surfaces transform irreversibly into the stable configurations upon increasing the temperature.)

While the preceding classification is certainly somewhat crude and does not include all the phenomena observed with ordered adsorbed phases, it suffices to allow the underlying principles to be recognized in all the following examples of real adsorption systems.

11.4. Physical adsorption

Physical adsorption denotes the weak interaction of gases with surfaces as a result of van der Waals forces without the formation of true chemical bonds i. e. specific interactions between the valence electrons of the substrate and the adsorbate. The small adsorption energy ($<$ 10 kcal/mole) involves the use of low temperatures and relatively high gas pressures in order to maintain appreciable surface coverages in stationary adsorption-desorption equilibria. These factors complicate experiments with clean surfaces under UHV conditions, since condensation of gaseous impurities may easily take place on the cooled surface. Furthermore the effect of the electron beam on the adsorbate is likely to be greater than in the case of tightly bound chemisorption complexes. These appear to be the main reasons for the fact that only a very limited number of investigations with well-defined single crystal surfaces has so far been performed, despite the considerable theoretical and practical interest in physisorption phenomena [7], [29]. Lander and Morrison [24] investigated the physical adsorption of various substances on basal planes of graphite using LEED and they observed the formation of a series of ordered adsorbed structures as well as different phase transitions. Adsorption of Xe causes the appearance of a dense hexagonal arrangement in registry with the substrate structure. An individual Xe atom covers an area of 15.7 Å² in this phase which is somewhat smaller than the value of 16.8 Å² in the corresponding densely packed plane of a solid Xe crystal. This result is of importance for methods of area determination based on physical adsorption, since a certain value of the "size" of the adsorbed particles has to be assumed.

A similar result was obtained in the case of adsorption of argon on a Nb(100) surface, where Dickey et al. [30] observed the formation of a hexagonal (111) plane of Ar atoms stacked on the substrate with the closed packed (110) rows lying in troughs formed by the rows of Nb atoms. The distance between rows of Nb atoms is 3.29 Å which is almost identical with the distance between close packed rows in solid argon (= 3.32 Å), so that as with Xe on graphite the overlayer structure is adjusted to that of the substrate by a minor variation of the interatomic distance.

Steiger et al. [31] investigated the physical adsorption of noble gases and hydrocarbons at (111) and (100) Ag surfaces by means of LEED and ellipsometry and were not able to detect the formation of ordered adsorbed phases in any case. However these results cannot be considered representative of the behaviour of clean surfaces since the adsorption of impurities such as water from the gas phase could not be ruled out [32].

The most extensive study so far was performed by Palmberg [33] with the system Xe/Pd(100). LEED observations revealed that the adsorbed layer is disordered up to high coverages. In the region of saturation the formation of a diffuse ring pattern indicated the existence of short-range order domains which transform at maximum coverage into a close packed hexagonal structure as shown in fig. 11.7. The adsorbed Xe atoms have a mutual distance of 4.48 Å (corresponding to a density of $5.8 \cdot 10^{14}$ cm^{-2}), a value which again differs only slightly from the bulk value of 4.37 Å. The kinetics of the formation of the adsorbed layer was followed by measuring the variation of the work function, of the Xe Auger intensity and of the LEED

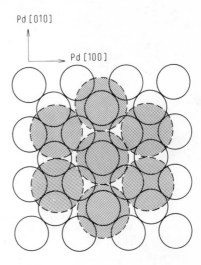

Fig. 11.7. Close packed hexagonal layer of Xe atoms adsorbed on Pd(100). After Palmberg [33].

'extra'-spot intensity (fig. 11.8). It was found that the sticking coefficient remains close to unity up to high coverages. The isosteric heat of adsorption E_{ad} was evaluated from measured adsorption isotherms (via $\Delta\phi$) using the Clausius-Clapeyron equation. At zero coverage E_{ad} has a value of 7.5 kcal/mole and decreases monotonically with increasing coverage due to dipole-dipole interactions. The dipole moment of an adsorbate complex was determined to be $^{\circ}\mu_s = 0.95$ Debye.

Similar observations of the structure of adsorbed Xe layers were made by Chesters and Pritchard [34] for Cu(100). Again a hexagonal arrangement was found with a Xe-Xe distance of 4.50 Å.

The general conclusion therefore is that the structure of the substrate has some influence on the orientation of the adsorbed noble gas layer, but that the configuration of the adsorbed atoms with respect to each other is mainly governed by their mutual interaction potentials which are only slightly influenced by the substrate. As a result the formation of close-packed hexagonal layers is observed in all cases. These findings justify the general assumptions underlying the usual models of physical adsorption, in which a certain area is attributed to an adsorbed particle irrespective of the structure and nature of the substrate.

11.5. Metallic adsorbates

The adsorption of metal atoms on surfaces is of considerable practical interest in projects such as thermionic energy conversion, nuclear reactor technology, crystal growth by epitaxy or chemical vapor deposition, thermal electron emission etc., and has already been studied

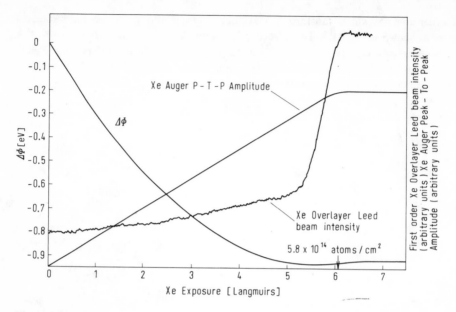

Fig. 11.8. Xe adsorption on Pd(100). Variation of the work function $\Delta\phi$, the Xe Auger-peak-to-peak amplitude, and the Xe overlayer LEED beam intensity with the Xe exposure at 77 K. After Palmberg [33].

extensively with various metal combinations. Some of the general aspects have been discussed by Moesta [35]. The most interesting observations have been made with alkali metals adsorbed on transition metal surfaces, and some of the major results from this class of systems are described in some detail below. These systems are characterized by a large decrease of the work function (see e.g. fig. 8.8), which is ascribed to an electron transfer from the adsorbed atoms (with small ionization energy) to the metal substrate. The binding is therefore predominately ionic, but for a complete description covalent contributions must also be taken into account [36]. Various theoretical descriptions are to be found in the literature [37].

A series of brilliant papers by Gerlach and Rhodin [38] concerns the formation of sodium overlayers on Ni single crystal planes. Although no attempt was made to analyse the LEED intensities, the structure models derived from the LEED patterns are highly probable (except the determination of the exact positions of the Na atoms with respect to the Ni atoms at low coverages), mainly because these results have been correlated with absolute determinations of the coverage.

Na adsorption on Ni(100) is characterized at low coverages by diffuse rings in the LEED pattern, which suggest the existence of uniformly spaced adatoms (i.e. only small standard deviations in their mutual average distance which of course varies continuously with θ) in registry with the substrate. The first monolayer is complete at $\theta = 0.5$ which is characterized by an ordered $c2 \times 2$-structure, corresponding to dense packing. In some aspects similar properties were exhibited by the Ni(111) surface. Again diffuse diffraction rings at low coverage indicate uniform spacing of the adatoms. At $\theta = 0.33$ a hexagonal structure is formed which is compressed continuously at increasing coverage until at $\theta = 0.49$ the first monolayer is completed. The final state is a hexagonal array of adatoms incoherently situated on the surface but with the same orientation.

The most interesting features were found with Na/Ni(110): At low coverage the formation of different structures again in registry with the substrate atoms is concluded. At $\theta = 0.25$ each other row of Ni atoms in the [100] direction is occupied by a chain of Na atoms with mutual distance $2a_1$ in [110] direction. These chains have discrete random shifts with respect to each other (fig. 11.9a). At $0.25 < \theta < 0.3$ the Na-Na spacing in [110] direction is continuously

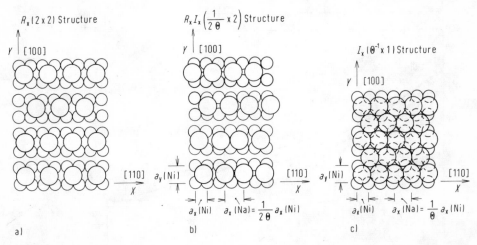

Fig. 11.9. Structure models for Na adsorbed on Ni(110). a) $\theta = 0.25$, b) $0.25 < \theta < 0.31$, c) $0.64 < \theta < 0.71$. After Gerlach and Rhodin [38b].

compressed (fig. 11.9b) leading to one-dimensionally incoherent structures. Up to $\theta = 0.64$ this Na-Na spacing in the [110] direction remains constant, but the remaining open troughs are successively filled until at $\theta = 0.64$ every trough contains an incoherent row of Na atoms, giving a distorted hexagonal structure (fig. 11.9c).

Finally some further compression in the [110] direction takes place until saturation of the first layer is reached at $\theta = 0.71$.

Maximum coverage in all three cases is characterized by a Na-Na distance which is about 8% smaller than that in metallic sodium and which can be ascribed to the partial electron transfer from the overlayer to the substrate. In the case of the Ni(100) surface the c2×2-structure exhibits the optimum Na-Na spacing thus preventing the formation of incoherent structures as is the case with Ni(110) and Ni(111). The competition between the trend towards an optimum surface coordination (=coherent structure) and close packing as determined by the adatom size (=incoherent structure) is nicely demonstrated in all stages of the various surface structures observed.

The work function decreases strongly and passes through a minimum for Na on all three planes (see fig. 8.8), as is generally observed with adsorption systems of this type, but the dipole moment of the adsorbate complexes calculated from $\mu_s = \Delta\phi/2\pi n_s$ (n_s = density of adsorbed atoms) varies continuously with increasing coverage without any peculiarities at the minima of $\Delta\phi$. Furthermore in no case is $\Delta\phi_{min}$ associated with any characteristic changes in the surface structures. It is remarkable that the (111) and (100) planes behave almost identically whereas the dipole moment at zero coverage is much smaller for Na/Ni (110). This may be due to the adsorbed atoms being placed in the troughs of the open (110) face. The binding energies were determined from thermal desorption data and are reproduced in fig. 11.10. Again the values for Ni(111) and (100) are nearly identical, whereas the initial

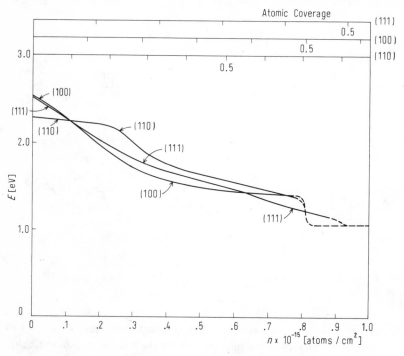

Fig. 11.10. Desorption energies for Na on Ni single crystal surfaces as function of the coverage. After Gerlach and Rhodin [38c].

desorption energy on Ni(110) is about 10% smaller. The thermal desorption spectra for Ni(100) and (111) are smooth curves and correlate with the observed continuous transformations of the surface structures. On the other hand various peaks appeared in the desorption spectra for Ni(110) which agree with the observed significant changes in the surface structures. It is important however to notice that each degree of coverage on Ni(110) is associated with a uniform adsorption structure, which means the occurrence of structure in the desorption spectra is not caused by the coexistence of several adsorbed states but by a continuous transformation of the arrangement of adsorbed particles during the desorption process.

Qualitatively very similar observations as with adsorbed Na were made for K and Cs adsorption on Ni(110)[38], but the larger size of the adatoms cause some differences. Again one-dimensional incoherent structures were observed for $\theta > 0.25$ indicating that the troughs in [110] directions present a rather smooth potential.

The atomic arrangement in the (112) surface of a bcc crystal corresponds to that of a fcc (110) plane. Chen and Papageorgopoulos [39] arrived at very similar conclusions for the structures of Na overlayers on W(112) as did Gerlach and Rhodin for Na/Ni(110). Again a coherent structure (2×1) forms at low coverage. At $\theta > 0.5$ the Na atoms are continuously compressed in the [110] direction, and again the minimum Na-Na spacing is somewhat smaller than in bulk sodium.

The formation of close-packed hexagonal layers at saturation of the first monolayer was also observed for the adsorption of Cs on Ni(111) [40], on W(110) [41], [42] and on W(100) [43]. Again the minimum spacing between adatoms was somewhat smaller than in bulk crystals. Interesting order-disorder phenomena have been observed by Fedorus and Nauvomets [41] with the Cs/W(110) system. At 300 K long-range order is lost in all layers with $\theta < 0.64$, whereas at 77 K layers with $\theta > 0.4$ are ordered and those with lower coverage disordered. The critical temperature for order-disorder transitions increases with increasing coverage which is in complete agreement with the assumption that the repulsion between neighbouring adatoms decreases with increasing average distance, i.e. decreasing coverage. The general picture which has been found to be valid for all alkali metal overlayers studied so far is not restricted to this class of adsorbates. Nauvomets and coworkers [43] studied the adsorption of Ba on Mo (110) and arrived at similar conclusions. A whole series of coherent adsorbate structures form at lower coverages, whereas a gradual compression of a $c2 \times 2$ unit cell takes place at higher θ until a hexagonal structure is formed which however remains in registry with the substrate in the [01̄1] direction, i.e. it is incoherent in one dimension. Finally this hexagonal structure is symmetrically compressed to a close packed layer. Similar order-disorder transitions were observed as with the system Cs/W(110), the critical temperatures being below room temperature and again decrease with decreasing coverage. It is interesting to notice that the work function change is only slightly affected by the occurrence of long-range disorder, probably because the short range order is not destroyed, which is the essential factor in determining the properties of the local electric potential barrier.

Many other metallic adsorbates have been studied, in particular in connection with problems of epitaxy. Some of these results were reviewed recently by Bauer and Poppa [44].

11.6. Chemisorption of gases

The chemisorption of small molecules such as CO, H_2 and O_2 has been the subject of even more investigations. CO chemisorption on metals of groups VIII and Ib is one of the best understood systems and can be used to illustrate the principles underlying this sort of work. A review of earlier investigations of CO adsorption on transition metals has been made by Ford [45]. It is generally accepted that the CO molecule is attached to the metal surface via the carbon atom. Observations of frequency shifts in the C-O stretching vibration of the adsorbed molecule frequently lead to a belief in the existence of two types of adsorbed complexes, 'linear' and 'bridged' by analogy with the formation of linear and bridged metal carbonyls. However as early as 1964 Blyholder [46] showed by a simple LCAO treatment that it was not necessary to consider different geometric configurations in order to explain the IR spectra.

Park and Madden [47] investigated the adsorption of CO on Pd(100) and observed the formation of two domain orientations of a $c(4 \times 2)/45°$-structure. They proposed a structure model as shown in fig. 11.11 which appears to be highly probable, since this is the only arrangement where using the unit cell obtained from the LEED pattern, all adsorbed CO molecules are

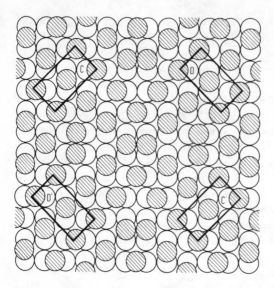

Fig. 11.11. Structure model for CO adsorbed on Pd(100), including the antiphase domains of the two orientations of the overlayer unit cell. After Park and Madden [47].

identically coordinated. Obviously this model corresponds to the 'bridge'-bond, an idea which is further supported by IR spectroscopy measurements with polycrystalline Pd samples [48]. However shortly after it was discovered [49] that this configuration does not represent saturation, but that the coverage may be further increased accompanied by a uniaxial continuous compression of the unit cell in the [100] direction, while the dimensions normal to the compression direction remain constant. The ordered c4 × 2/45°-structure forms just below $\theta = 0.5$. The conclusion is that short range repulsive interactions between the adsorbed CO molecules lead to a random lattice gas at $\theta < 0.5$. The situation differs somewhat from the Na/Ni (100) system [38] in so far as there the action of long range dipole-dipole interactions leads to uniform spacing between the adatoms as indicated by the appearance of ring patterns at low surface concentrations. The isosteric heat of CO adsorption on Pd(100) E_{ad} as a function of the coverage decreases continuously up to $\theta = 1/2$ due to the mutual repulsion between adsorbed particles. The onset of compression, i.e. the loss of symmetric adsorption sites, is accompanied by a steep decrease of E_{ad} by about 7 kcal/mole [49a].

Similar observations were made with CO/Pd(111) [50]. At $\theta = 1/3$ a $\sqrt{3} \times \sqrt{3}/30°$-structure is fully developed. But, in contrast to Pd(100), Pd(110) and Pd(210), sharp but weak LEED 'extra' spots became visible at lower coverage which must be ascribed to attractive interactions between adsorbed CO molecules leading to island formation. It is assumed that this attraction is caused by indirect interactions [23] which may depend on the crystallographic orientation of the surface. Again the unit cell of this coherent structure is continuously compressed upon increasing the coverage beyond $\theta = 1/3$ and a dense packing of adsorbed particles is achieved. Fig. 11.12 shows the isosteric heat of adsorption E_{ad} for this system. Due to the attractive interactions at low coverage E_{ad} remains constant up to $\theta = 1/3$. The onset of the uniaxial compression is accompanied by a sudden decrease of E_{ad} by only 2 kcal/mole. Obviously this plane is energetically smoother than the (100) surface.

It is interesting to compare the initial*) heats of CO adsorption on different single crystal

---

*) i.e. the values extrapolated to zero coverage.

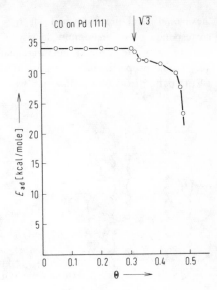

Fig. 11.12. Isosteric heat of CO adsorption on Pd(111) as a function of coverage θ. After [50].

planes. Table 11.1 shows that these differences do not exceed about 15% of the binding energy, which means that no pronounced plane specifity for this adsorption system exists [51]. (These results may also be compared with the system Na/Ni).

Table 11.1. Initial heats of CO adsorption on different Pd single crystal surface [51].

Surface	(111)	(100)	(110)	(210)	(311)
E_{ad}° [kcal/mole]	34	36.5	40	35.5	35

The idea that the geometric structure of the surface is not of much significance in the overall adsorption properties of metals is further supported by studies with CO adsorbed on nickel. The system CO/Ni has been studied very extensively in the past using both single crystals and polycrystalline samples, but it is only recently that a more or less complete understanding has become possible. Previous studies were frequently influenced by surface impurities and by the tendency of CO adsorbed on nickel to decompose into oxygen and carbon under the influence of an electron beam or thermal treatment [10c, 52].

A careful study with Ni(100) was performed by Tracy[52]. From the LEED patterns the existence of several ordered and disordered structures was deduced. A c2×2-structure corresponds to a coverage $\theta = 1/2$ for which the adsorbed CO molecules are situated on those Ni 'adsorption sites' with fourfold symmetry, i.e. either on top of the Ni atoms or between 4 surface atoms. (This fact was deduced from characteristic broadening of some of the LEED spots at partial disorder). The structure model of fig. 11.13 prefers the fourfold coordination. Adjacent adsorption sites cannot be occupied due to the size of the CO molecule. At $\theta = 0.61$ a distorted 'hexagonal' structure is formed which is further continuously compressed up to $\theta = 0.69$, corresponding to a maximum density of adsorbed CO molecules of 1.10×10^{15} cm^{-2}. In addition interesting observations were made of disordered and two-phase regions.

Very similar properties have been found with CO adsorbed on Ni(110) [53], [54]. At $\theta = 0.7$ a LEED pattern was observed which was interpreted as being caused by a one-dimensional

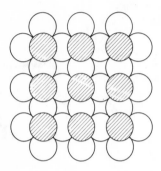

Fig. 11.13. Structure model for CO adsorbed on Ni(100) at $\theta = \frac{1}{2}$. After Tracy [52].

incoherent structure as shown in fig. 11.14a, where the CO molecules fill the troughs on Ni(110) [53]. (A somewhat different model for this structure was proposed by Taylor and Estrup [54]). At saturation a 2×1-structure is formed (fig. 11.14b) which corresponds to a surface density of 1.12×10^{15} cm^{-2}, i.e. exactly the same value as for CO/Ni(100). This finding is in complete agreement with measurements of Klier et al. [55] who used radiotracer techniques to measure the adsorbed amount and found $n_s = 1.11 \times 10^{15}$ cm^{-2}. The isosteric heat of CO adsorption on Ni(110) is 30 kcal/mole up to about $\theta = 0.6$ after which it decreases to a value of 25 kcal/mole [53]. The same initial value was also found for Ni(100) [52]. Obviously again the structural factor is only of minor importance for this system. This is confirmed by measurements with polycrystalline Ni films which agree very well with the single crystal data [56].

With Cu(100) the formation of a c2 × 2-structure was observed after CO adsorption which then became continuously compressed in the [110] direction with further increase of the coverage [57]. The maximum coverage was about 0.9×10^{15} cm^{-2} which is somewhat smaller than on Ni(100) and on Pd(100). The reason is that the heat of adsorption on Cu is only about 15 kcal/mole as compared with 30 and 36 on Ni and Pd, respectively, and there is therefore less energy available to pack the CO molecules to higher density against their mutual repulsive barrier. The general conclusion to be drawn from a comparison of the systems discussed above is that the localization of the adsorbed CO molecules on specific high coordination substrate sites is not very critical. Furthermore the crystallographic orientation of the surface is of minor importance with these systems, which is an advantage when looking at the electronic aspects of CO chemisorption which should therefore be mainly governed by the bulk electronic properties.

In contrast to the 'structural' effect the 'electronic' factor is very important. For example CO adsorption causes an increase of the work function on Ni and a decrease on Cu [57b] and the adsorption energies differ considerably. An analogous situation occurs for Ag and Pd, whereas Ni and Pd have quite similar properties for CO adsorption.

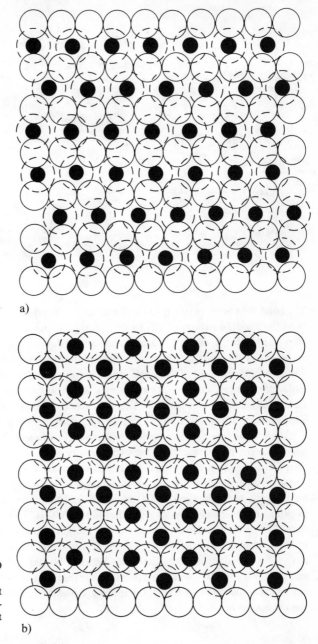

Fig. 11.14. Structure models for CO adsorbed on Ni(110).
a) One-dimensionally incoherent structure at $\theta \approx 0.7$. b) c(2 × 1)-structure at $\theta = 1$. After Madden et al. [53].

Studies with Cu/Ni [58] and Ag/Pd alloys [59] revealed that no linear relationship between alloy composition and adsorption properties exists for these systems. Fig. 11.15 shows the maximum change of work function $\Delta \phi$ after CO adsorption as a function of the surface composition of Ag/Pd alloys (as determined by AES) [59].

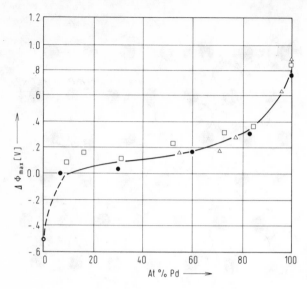

Fig. 11.15. Maximum change of the work function $\Delta\phi_{max}$ after CO adsorption on Ag/Pd alloys as a function of the alloy surface composition. After [59].

The bond formation between CO and a metal surface takes place through an electron transfer from the filled 3σ orbital of CO to the metal and back-donating of metallic electrons to the empty $2\pi^*$ states. Using this concept Grimley [60] developed a model theory for CO adsorption on Ni which accounts for the order of magnitude of the heat of adsorption and predicts the direction of the change of work function correctly. The variation of the occupancy of the molecular orbitals changes their energies which are broadened through the formation of virtual resonant states. This prediction is in good agreement with the UPS results of Eastman and Cashion [61] as discussed in section 4.5, where a shift and broadening of the 3σ orbital was found. The energy of the higher filled orbital involving the CO-$2\pi^*$ state should be located just below the Fermi level. The emitted intensity is observed to decrease in this energy range after CO adsorption which is probably due to this effect.

The main contribution to the adsorption energy arises from the transfer of d-electrons from the metal to the $2\pi^*$-orbital of the CO molecule, and therefore the higher on the energy scale the position of the d-band is the stronger the binding [60], [62]. On the other hand electron transfer to the $2\pi^*$-level should lead to an increase of the work function and to a lowering of the strength of the C-O bond and therefore of the C-O stretching frequency in accordance with the IR data. This explains also the differences between Ni, Pd and Cu, Ag. The alloy experiments and further theoretical treatments [62] indicate that not the individual surface atoms but rather ensembles consisting of a few metal atoms are essentially responsible for the formation of the chemisorption bond.

11.7. Co-adsorption

If two different kinds of particles 1 and 2 are adsorbed on a surface certain new aspects come into play. In contrast to single component systems the structure of the adsorbed layer is then determined by three types of interactions between adsorbed particles ε_{11}, ε_{22} and ε_{12} (assuming only pairwise interactions between nearest neighbours). Further complications may arise in so far as the sequence of adsorption may be of primary importance, allowing either displacement reactions to take place, or the establishment of thermodynamic equilibrium to be inhibited by kinetic phenomena, such as nucleation etc.

As regards the arrangement of the adsorbed particles with respect to each other the situation is similar to that in ordinary mixed phase thermodynamics. If entropy effects are neglected (i.e. at absolute zero of temperature) the nature of intermolecular interactions leads either to the formation of ordered mixed phases or to the occurrence of a complete miscibility gap. At $T > 0$ intermediate stages between these two extreme cases occur, e.g. partial miscibility and disordered structures. If these principles are applied to adsorbed layers again two limiting situations have to be discussed which depend on the intermolecular interactions [63]:

a) The formation of a regular mixed phase as illustrated schematically in fig. 11.16 if $\varepsilon = \varepsilon_{11} + \varepsilon_{22} - 2\varepsilon_{12} < 0$ (or roughly speaking if the interaction between 1 and 2 is more attractive than that between identical particles). This situation may be called *cooperative adsorption*. Both kinds of particles are in intimate contact and owing to the altered interaction between neighbouring particles a new periodicity is formed which can easily be detected in the LEED pattern.

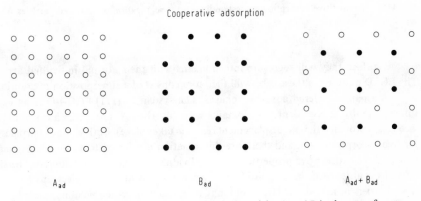

Fig. 11.16. Scheme of the configurations of adsorbed particles A and B in the case of cooperative adsorption.

For example CO forms a c2 × 2-structure on a Pd(110) surface at moderate coverage whereas hydrogen causes a 1 × 2-structure [64]. If a partially CO-covered Pd (110) surface is exposed to H_2 a 1 × 3 structure appears which must be ascribed to a mixed adsorbate complex. The sequence of adsorption as well as the initial CO coverage is very important in this case; if a hydrogen covered surface is exposed to CO, the latter may completely displace the H-adlayer. Obviously the change in intermolecular interaction and surface periodicity can also lead to variations in the adsorption energy, as reported for example for the systems W(100) + N_2 + CO [65] and Pt(110) + S + CO [66]. These effects are roughly an order of magnitude smaller than the adsorption energies. The modification of infrared band positions for adsorbed CO by the presence of NH_3 as described by Blyholder and Sheets [67] is also to be ascribed to a similar cooperative adsorption effect leading to electronic interactions. It is further evident that the total amount adsorbed and other surface properties such as the change in the work function need not be a simple addition of the values due to the presence of the individual adsorbates. It is felt that cooperative adsorption is favourable for catalytic reactions since the reactants are in intimate contact and a high frequency factor in the rate constant should result.

b) The other extreme situation occurs if the particles of type 1 and 2 are completely immiscible, for example if they repel each other strongly (fig. 11.17). If both kinds of particles are present

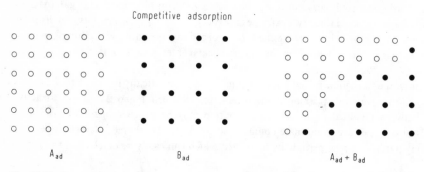

Fig. 11.17. Competitive adsorption corresponding to the segregation of adsorbed particles A and B into domains.

on the surface they will segregate into domains with their normal individual arrangement. The LEED patterns in these cases will therefore consist of a superposition of additional spots of both separate ordered adsorbed phases. The system $Pd(111) + CO + O$ is an example of this type [50b]. It is evident that domains with adsorbed A inhibit the adsorption of B and vice versa (provided that no displacement reaction takes place). Both kinds of particles compete for free adsorption sites, and therefore this situation is referred to as *competitive adsorption*. Obviously the adsorptive properties of the individual particles are not altered. The size of the domains is at least of the magnitude of the coherence width of the electrons (~ 100 Å), but can be orders of magnitude larger and may even reach macroscopic dimensions [66].

It is not unreasonable to assume that competitive adsorption is not so favourable for catalytic reactions, since the reactants are only in contact at the boundaries of the domains ('boundary reaction') and thus the pre-exponential factor in the rate law is small. But even more important is the fact that in such cases the adsorption of one reactant may completely inhibit the adsorption of the other eventually leading to a complete change of the rate-determining step in steady-state catalysis.

11.8. Reactions at surfaces

Chemical reactions in adsorbed layers form the essential steps in heterogeneous catalysis. Reactions between adsorbed particles and the substrate leading to bulk compounds (e.g. oxides), volatile products (like GeO or GeH$_4$) or crystal growth (e.g. in chemical vapour deposition) are also fundamental chemical processes at solid surfaces, but are not strictly two-dimensional phenomena and are therefore excluded from the following discussion.

Catalysis denotes the acceleration of a chemical reaction by means of a substance which is not consumed under steady-state conditions. The main function of a catalyst is to form intermediate compounds with the reactants (i.e. adsorbate complexes in heterogeneous catalysis) thereby offering an alternative reaction path which allows a higher rate of reaction. Very frequently the reaction products remain adsorbed on the surface with an activation energy barrier to surmount before desorption. Since only a limited number of sites is available on the surface the interplay between the different kinds of adsorbed particles will determine their net rate of reaction under steady-state conditions. A detailed investigation of the individual steps is therefore indispensible when studying the reaction mechanism. Again only a few examples can be discussed here. More information on aspects of the application of LEED to catalytic problems can be found in two review papers [68]. Two groups of surface reactions may be considered, decomposition (i.e. A → products) and synthesis (i.e. A + B → products). Decomposition reactions frequently occur via several intermediate steps. A good example is the interaction of ethylene with Ni(110), where a stepwise dehydrogenation after thermal treatment of the adsorbate covered surface leads to the formation of an ordered 5×4 carbon overlayer which transforms finally into characteristic ring-like LEED patterns which are attributed to graphite [63].

A similar picture was obtained from studies of ammonia decomposition on W(100) [69] and W(112) [70]. At room temperature NH_3 adsorbs on both planes without forming ordered structures. At 800 K part of the hydrogen desorbs (detected by mass spectrometry) and ordered structures appear which are ascribed to adsorbed NH_2 radicals. At even higher temperatures nitrogen and hydrogen desorb via a series of complex intermediate structures.

Dehydrocyclization of n-heptane on Pt single crystals was studied by Joyner et al. [71] and is another example of this type of decomposition reaction which can be studied under UHV conditions. In this work special precautions for preventing spurious reactions in the UHV chamber were taken by gold-plating the walls. The formation of toluene was detected by a mass spectrometer. The activities of surface steps which were introduced by the use of very high index planes were of particular interest in this investigation. Among other features these steps appeared to be important in the nucleation of ordered carbon structures during the dissociative chemisorption of the reactants.

Synthesis reactions may take place by two basic mechanisms: a) The Langmuir-Hinshelwood mechanism, where both components are adsorbed and react in the adsorbed state. b) The Eley-Rideal mechanism, where only one reactant is chemisorbed and the other reacts from the gas phase or from a weakly bound physisorbed layer. Reactions of this type which have been studied so far concern typically the removal of an adsorbed species from the surface by interaction with another ('clean-off' reactions).

Germer and May [72] studied the reduction of an oxygen covered Ni(110) surface by hydrogen using visual examination of the oxygen LEED spots to monitor the clean-off. A similar more quantitative procedure was used to study the interaction between adsorbed oxygen and gaseous CO on Cu(110) [73]. In this case oxygen forms two ordered structures and the intensities of the corresponding LEED spots were measured as a function of CO exposure and of temperature. It was concluded that the reaction starts at 'active centres' (probably structural defects). This idea has been extended and discussed theoretically by May [74] who correlated the junctions of domain boundaries in adsorbed layers with the enhanced activity. As a consequence clean-off reactions may start at surface defect sites followed by growth around these sites of 'holes' leading to an autocatalytic effect, i.e. a speeding up of the rate of reaction at intermediate surface concentrations. Such an effect was also postulated by Hudson and Holloway [75] for the interaction between oxygen and sulfur on Ni(111). Sulfur forms two ordered structures on this plane and may be removed by O_2 even at room temperature forming SO_2. Fig. 11.18

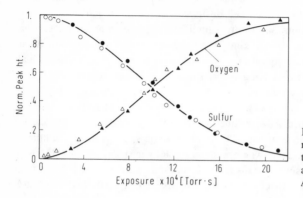

Fig. 11.18. Fractional coverage, measured by the Auger peak-to-peak amplitudes, of oxygen and sulfur on Ni(111) as a function of O_2 exposure at 30 °C. After Holloway and Hudson [75].

shows the fractional surface coverage of oxygen and sulfur as a function of O_2 exposure measured by AES. The reaction rate is proportional to the oxygen pressure and does not depend on the concentration of adsorbed oxygen, and a mechanism involving physisorbed O_2 was proposed. Bonzel [76] studied the same reaction at Cu(110) following the surface concentration with AES but at much higher temperatures and somewhat different conclusions were reached.

The occurrence of the type of boundary reaction under discussion was demonstrated very nicely by Bonzel and Ku [66] for the reaction between sulfur and oxygen on Pt(110). It was found that the rate of SO_2 formation decreased with increasing initial sulfur coverage θ_s leading to considerable 'incubation' times at high θ_s. Using the scanning AES technique the sulfur distribution on the surface was measured after partial reaction (see fig. 2.5). The result was that with high initial S coverages the reaction was nucleated at several locations on the surface, predominantly near the edges, leading to the formation of sulfur islands with up to 1.5 mm diameter. The surface diffusion coefficient of adsorbed sulfur was estimated to be about $3 \cdot 10^{-7}$ cm^2/s at 450 °C.

The oxidation of carbon monoxide at Pd surfaces represents one of the best understood systems at present [77]. Oxygen adsorbs on different single crystal planes of Pd with a mean heat of adsorption of about 60 kcal/mole and desorbs with second order kinetics which is consistent with the adsorbed state being composed of atomic oxygen. CO adsorbs without dissociation with a mean adsorption energy of 35 kcal/mole. CO_2 is formed via an interaction of the lone electron pair at the C atom of CO with the p_x-orbital of O forming a σ-bond, and through overlap of p_y-orbitals of C and O giving a π-bond. CO_2 is only very weakly bound to the surface and desorbs immediately above room temperature. Three basic reaction schemes are possible:

a) $O_{ad} + CO \rightarrow CO_2$ (Eley-Rideal reaction). This reaction should take place readily, since the lone electron pair at the C atom of CO is completely free for interaction with O_{ad}. In fact exposing an oxygen covered Pd surface to gaseous CO leads to an extremely rapid formation of CO_2 even at room temperature, as can be seen from changes in the LEED pattern [50], [77].

b) $CO_{ad} + 1/2\, O_2 \rightarrow CO_2$ (Eley-Rideal reaction). This reaction seems to be very unfavourable, since the O-O bond is unaffected by the catalyst and the active orbitals of CO are involved in the chemisorption bond. Indeed a surface completely covered with adsorbed CO is totally spectrometric evidence.

c) $O_{ad} + CO_{ad} \rightarrow CO_2$ (Langmuir-Hinshelwood reaction). If a Pd surface only partially covered by adsorbed CO is exposed to oxygen this adsorbs on the bare sites and two-phase competitive adsorption occurs, giving a superposition of O- and CO- 'extra' spots in the LEED pattern [50]. In this situation O_{ad} and CO_{ad} can only interact at the boundaries of the domains. Such two-phase surfaces are, not surprisingly, rather stable at room temperature, the formation of CO_2 only taking place after the temperature is raised to about 80 °C. Very similar conclusions were reached by Bonzel and Ku [78] for the same reaction on a Pt(110) surface, mainly from mass spectrometric evidence.

It is concluded that the overall reaction is mainly governed by the reaction a) and that therefore the rate of CO_2 formation is given by $r \approx k \cdot [O_{ad}] \cdot p_{CO}$, where $[O_{ad}]$ is the surface concentration of adsorbed oxygen. It must be remembered however that catalysis is normally a steady-state process. Competition takes place between O_2 and CO for free adsorption sites, and it is evident that at lower temperatures the surface will be completely covered by adsorbed CO and the reaction can only start at temperatures where CO desorption sets in. This can be seen from fig. 11.19 where the steady-state rate of CO_2 for-

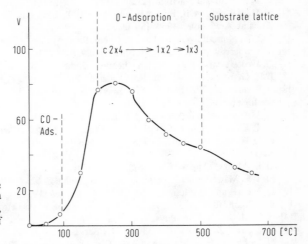

Fig. 11.19. Steady-state rate of CO_2 formation at a Pd(110) surface in a CO/O_2 mixture ($p_{O_2} = 1 \times 10^{-7}$ Torr, $p_{CO} = 1.6 \times 10^{-7}$ Torr) as a function of temperature [77b].

mation on Pd(110) in a constant CO/O_2 mixture is plotted as a function of temperature, while the state of adsorption was simultaneously monitored by LEED [77b]. At temperatures below 200 °C desorption of CO is rate-determining leading to an activation energy of about equal to that of CO desorption. At $200 < T < 300$ °C the maximum coverage of O_{ad} is reached and owing to the low activation energy for the reaction $O_{ad} + CO \rightarrow CO_2$ the rate is nearly independent of temperature. At still higher temperatures oxygen starts to desorb, the stationary concentration of O_{ad} is lowered (as can be seen from the gradual degradation of the LEED 'extra' spots), and the reaction rate decreases.

There are several factors which influence the effectiveness of a catalyst, by far the most important of which is poisoning.

Bonzel and Ku [66] investigated the influence of a sulfur overlayer on a Pt(110) surface on the steady-state rate of CO_2 formation. It was found that the activity decreased strongly with increasing S coverage θ_s (monitored by AES) and disappeared completely at $\theta_s = 0.3$. Although CO adsorption is still possible at $\theta_s = 0.3$ the reaction time for the removal of sulfur by reaction with oxygen increases very steeply, so that the sulfur on the surface inhibits the adsorption of oxygen, thus hindering the oxidation reaction. On the other hand if a Pd catalyst surface contaminated by C and S is treated for some time in a CO/O_2 mixture at elevated temperatures the activity gradually increases and finally becomes reproducible. It was shown by AES that the impurities were removed from the surface by reaction with oxygen [77a]. Such 'break-in' phenomena are well known in real catalysis and the above mentioned example shows that it is quite possible that a 'real' catalytic reaction proceeds on essentially clean surfaces which are formed during the reaction.

11.9. References

[1] J. H. Block and M. S. Zei, Surface Sci. *27*, 419 (1971).
[2] A. Benninghoven, Surface Sci. *28*, 54 (1971); *35*, 427 (1973).
[3] a) G. Ehrlich, Advances Catalysis *14*, 255 (1963).
   b) P. A. Redhead, Vacuum *12*, 203 (1962).
[4] N. M. Bashara, A. B. Buckman and A. C. Hall, eds.: Recent developments in ellipsometry. North Holland, Amsterdam 1969. (Surface Sci. *16* (1969)).
[5] J. C. Tracy and P. W. Palmberg, J. Chem. Phys. *51*, 4852 (1969).
[6] M. Perdereau and J. Oudar, Surface Sci. *20*, 80 (1970).
[7] G. Wedler: Adsorption. Verlag Chemie, Weinheim 1970.
[8] L. H. Germer and A. U. MacRae, J. Chem. Phys. *37*, 1382 (1962).
[9] B. Boddenberg, R. Haul and G. Oppermann, Adv. Mol. Relaxation Processes *3*, 61 (1972).
[10] See for example:
   a) J. Anderson and P. J. Estrup, Surface Sci. *9*, 463 (1968).
   b) T. Edmonds and R. C. Pitkethly, Surface Sci. *15*, 137 (1969).
   c) H. H. Madden and G. Ertl, Surface Sci. *35*, 211 (1973).
[11] Reviews:
   a) D. Menzel, Angew. Chemie *82*, 263 (1970).
   b) T. E. Madey and J. T. Yates, J. Vac. Sci. Techn. *8*, 525 (1971).
[12] D. Menzel and R. Gomer, J. Chem. Phys. *40*, 1164 (1964); *41*, 3311, 3329 (1964).
[13] P. A. Redhead, Can. J. Phys. *42*, 886 (1964); Appl. Phys. Lett. *4*, 166 (1964).
[14] G. E. Moore, J. Appl. Phys. *32*, 1241 (1961).
[15] D. Menzel, Z. Naturf. *23a*, 330 (1968); Ber. Bunsenges. *72*, 591 (1968).
[16] D. Lichtman, F. N. Simon and T. R. Kirst, Surface Sci. *9*, 325 (1968).
[17] T. E. Madey and J. T. Yates, Surface Sci. *11*, 327 (1968).
[18] J. T. Yates and D. A. King, Surface Sci. *32*, 479 (1972).
[19] I. G. Newsham, J. V. Hogue and D. R. Sandstrom, J. Vac. Sci. Techn. *9*, 596 (1972).
[20] I. G. Newsham and D. R. Sandstrom, J. Vac. Sci. Techn. *10*, 39 (1973).
[21] P. Kronauer and D. Menzel, in: Adsorption-Desorption Phenomena (F. Ricca, ed.) Academic Press, New York 1972, p. 313.
[22] J. H. de Boer: The dynamical character of adsorption. Oxford press 1968.
[23] a) T. B. Grimley, Proc. Phys. Soc. (London) *90*, 75 (1967); *92*, 776 (1967).
   b) T. L. Einstein and J. R. Schrieffer, J. Vac. Sci. Techn. *9*, 956 (1972; Phys. Rev. *B 7*, 3629 (1973).
[24] J. J. Lander and J. Morrison, Surface Sci. *6*, 1 (1967).
[25] G. Ertl and J. Küppers, Surface Sci. *21*, 61 (1970).

[26] J. Küppers, M. Hollemann-Plancher and G. Ertl, Proc. Spring Meeting of the German Physical Society, Münster 1971 (unpublished).

[27] G. Doyen, Diplomarbeit, Technical University Hannover, 1972.

[28] P.J. Estrup, in: The structure and chemistry of solid surfaces (Ed. G.A. Somorjai), Wiley, New York 1969, p. 19-1.

[29] a) D.M. Young and A.D. Crowell: Physical Adsorption of Gases. Butterworths, London 1962.
     b) S. Ross and J.P. Olivier: On Physical Adsorption. Wiley, New York 1964.
     c) Adsorption-Desorption Phenomena (F. Ricca, ed.) Academic Press, New York 1972.

[30] J.M. Dickey, H.H. Farrell and M. Strongin, Surface Sci. *23*, 448 (1970).

[31] R.F. Steiger, J.M. Morabito, G.A. Somorjai and R.H. Muller, Surface Sci. *14*, 279 (1969).

[32] L.A. Bruce, Surface Sci. *20*, 187 (1970).

[33] P.W. Palmberg, Surface Sci. *25*, 598 (1971).

[34] M.A. Chesters and J. Pritchard, Surface Sci. *28*, 460 (1971).

[35] H. Moesta: Chemisorption und Ionisation in Metall-Metall-Systemen. Springer Verlag, Berlin 1968.

[36] L.D. Schmidt and R. Gomer, J. Chem. Phys. *42*, 3573 (1965); *45*, 1605 (1966).

[37] a) E.P. Gyftopoulos and J.D. Levine, Surface Sci. *1*, 171, 225 (1964).
     b) J.W. Gadzuk, Surface Sci. *6*, 133 (1967); *11*, 465 (1968).
     c) B.J. Thorpe, Surface Sci. *33*, 306 (1972).

[38] R.L. Gerlach and T.N. Rhodin,
     a) Surface Sci. *10*, 446 (1968);
     b) *17*, 32 (1968);
     c) *19*, 403 (1970).

[39] J.M. Chen and C.A. Papageorgopoulos, Surface Sci. *21*, 377 (1970).

[40] T. Calcott and A.U. MacRae, Phys. Rev. *178*, 966 (1969).

[41] A.G. Fedorus and A.G. Nauvomets, Surface Sci. *21*, 426 (1970).

[42] A.U. MacRae, K. Müller, J.J. Lander and J. Morrison, Surface Sci. *15*, 426 (1969). The original interpretation of the observed LEED patterns by a two-layer model proved to be incorrect and has been revised in the meantime (K. Müller, personal communication).

[43] A.G. Fedorus, A.G. Nauvomets and Yu.S. Veluda, Phys. Stat. Sol. *13a*, 445 (1972).

[44] E. Bauer and H. Poppa, Thin Solid Films *12*, 167 (1972).

[45] R.R. Ford, Adv. Catalysis *21*, 51 (1970).

[46] G. Blyholder, J. Phys. Chem. *68*, 2772 (1964).

[47] R.L. Park and H.H. Madden, Surface Sci. *11*, 188 (1968).

[48] a) N.N. Kavtaradze and N.D. Sokolova, Zh. Fiz. Khim. *41*, 225 (1967).
     b) J.K.A. Clarke, G. Farren and M.E. Rubalacava, J. Phys. Chem. *71*, 2376 (1967).

[49] a) J.C. Tracy and P.W. Palmberg, J. Chem. Phys. *51*, 4852 (1969).
     b) G. Ertl and J. Koch, Z. phys. Chem. N.F. (Frankfurt) *69*, 323 (1970).

[50] G. Ertl and J. Koch,
     a) Z. Naturforschung *25a*, 1906 (1970);
     b) in: Adsorption-Desorption Phenomena (F. Ricca, ed.) Academic Press, New York 1972, p. 345.

[51] H. Conrad, G. Ertl, J. Koch and E.E. Latta, Surface Sci. *43*, 462 (1974).

[52] J.C. Tracy, J. Chem. Phys. *56*, 2736 (1972).

[53] H.H. Madden, J. Küppers and G. Ertl, J. Chem. Phys. *58*, 3401 (1973).

[54] T.N. Taylor and P.J. Estrup, J. Vac. Sci. Techn. *10*, 26 (1973).

[55] K. Klier, A.C. Zettlemoyer and H. Leidheiser, J. Chem. Phys. *52*, 589 (1970).

[56] G. Wedler, H. Papp and G. Schroll, Surface Sci. (in press).

[57] a) J.C. Tracy, J. Chem. Phys. *56*, 2748 (1972).
     b) J. Pritchard, J. Vac. Sci. Techn. *9*, 895 (1972).

[58] G. Ertl and J. Küppers, J. Vac. Sci. Techn. *9*, 829 (1972).

[59] K. Christmann and G. Ertl, Surface Sci. *33*, 254 (1972).

[60] T.B. Grimley, in: Molecular Processes at Solid Surfaces. (Eds. Drauglis, Gretz, Jaffee) McGraw Hill, New York 1969, p. 299.

[61] D.E. Eastman and J.K. Cashion, Phys. Rev. Lett. *27*, 5120 (1971).

[62] G. Doyen and G. Ertl, Surface Sci. *43*, 197 (1974)

[63] G. Ertl in: Molecular Processes at Solid Surfaces. (Eds. Drauglis, Gretz, Jaffee), McGraw Hill, New York 1969, p. 147.

[64] H. Conrad, G. Ertl and E. E. Latta, Surface Sci. *41*, 435 (1974).
[65] P. J. Estrup and J. Anderson, J. Chem. Phys. *46*, 567 (1967).
[66] H. P. Bonzel and R. Ku, J. Chem. Phys. *59*, 1641 (1973).
[67] G. Blyholder and R. W. Sheets, J. Catalysis *27*, 301 (1972).
[68] a) J. W. May, Adv. Catalysis *21*, 151 (1970).
     b) G. Ertl, Proc Int. LEED Summer School Smoleniče (CSR), (ed. M. Lacnička), Union of Cze-
        choslovak Mathematicians and Physicists, Prague 1972, Vol. 1, p. 214.
[69] P. J. Estrup and J. Anderson, J. Chem. Phys. *49*, 523 (1968).
[70] J. W. May, R. J. Szostak and L. H. Germer, Surface Sci. *15*, 37 (1969).
[71] R. W. Joyner, B. Lang and G. A. Somorjai, J. Catalysis *27*, 505 (1972).
[72] L. H. Germer and J. W. May, in: The structure and chemistry of solid surfaces. (Ed. G. A. Somorjai)
     Wiley, New York 1969, p. 51-1.
[73] G. Ertl, Surface Sci. *7*, 309 (1967).
[74] J. W. May, Proc. Roy. Soc. (London), in press.
[75] P. H. Holloway and J. B. Hudson, Surface Sci. *33*, 56 (1972).
[76] H. P. Bonzel, Surface Sci. *27*, 387 (1971).
[77] G. Ertl and J. Koch, Proc. Vth Int. Congress on Catalysis, Palm Beach (USA), North Holland,
     Amsterdam 1973, p. 969.
     b) G. Ertl and P. Rau, Surface Sci. *15*, 443 (1969).
[78] H. P. Bonzel and R. Ku, Surface Sci. *33*, 91 1972.
[79] W. Heiland and E. Taglauer, J. Vac. Sci. Techn. *9*, 620 (1972).

Subject Index

adsorption-bond 79, 125, 194, 207, 212, 216, 230, 235
— co *230*–232
— competetive 138, 232, 235
— cooperative 231
— energy (→ heat-adsorption)
— isotherms 221
— kinetics 83, 212, 214, 221
— physical *220*–221
— state 81, 114, 207, 214f.
AEAPS (→ spectroscopy-AEAPS Auger electron appearance potential spectroscopy)
AEAPS-spectrometer 87
AES (→spectroscopy-AES-Auger electron spectroscopy)
alloys 6, *43*–45, *77*–79, 85, *93*–95, *185*, 229
analyzer-CHA *14*–15, 71
— CMA *12*–13, 15, 17, 21ff., 46f., 54
— dispersion 12f., 15, 70
— field electron energy *99*, 102, 105
— hemispherical *14*–15, 71
— multichannel 23, 71, 211
— RFA (retarding field) *9*–12, 17, 20ff., 46, 49, 54, 67, 71, 99
— spherical deflector (→ CHA)
— 127° *13*–14, 17, 19
analysis
— diffraction pattern *145*–146, 151f., 156, 171, 176, 178f., 181
— intensity 163, 171, 176, 178f., 181, 183, 186, 222
— micro 18
— qualitative *41*
— quantitative 19f., 39, *41*–47
— structural 9, 152, 181f.
— x-ray 134f., 140
angle diffraction 138, 167, 180, 182
— glancing 17ff., 22, 44, 87
— incidence 18, 21, 40f., 133, 198f.
— scattering 180, 195
angular distribution 38, 54, 61, 79, 129
— profile 61
annealing 5ff., 44, 129, 186
APS (→ spectroscopy-APS-Appearance potential spectroscopy)
APS-mechanism *88*–91
— peak-energy *91*–92
— peak height 91, 93, 97, 211
— peak-shape 91ff., 97
— peak-width 92, 94, 97
— spectrometer *86*–88
atomic ionization cross section 18, 37
attenuation factor 38, 175

Auger-electron 5, 9, *17*–51, 53, 55, 68f., 72, 85, 110
— electron-spectroscopy (→ spectroscopy-AES)
— intensity 22, *34*–39, 43, 220
— mechanism *24*–30
— peak-energy 11, 20, *30*–34, 41, 48
— peak-height 23, 34, 40ff., 234
— peak-shape 11, 20, *30*–34, 41, 48
— peak-width 26
— spectrometer 18, *19*–24, 31, 33, 68f.
— yield 18, 29, *34*–36, 85
averager-signal 23, 71
averaging-method (-technique) 152, 154, *179*–182

background 9ff., 17f., 20, 22f., 49, 62, 69, 87f., 93, 159, 164, 175, 185, 194, 202, 205
backscattering 8f., 18, 38ff., 53f., 60ff., 64, 72, 129, 133, 138, 140, 172f., 176, 178, 196f., 188
— coefficient 39
band-bending 72
— d 76, 78ff., 91ff., 97, 102, 113f., 230
— p 78ff., 91ff., 97, 102, 113f., 230
— s 78, 93, 113
— structure 29, 31, *47*–49, *67*–84, 85, *93*–95, 102, 174, 176f.
— surface 82
— valence · 29ff., 49, 53, 72, 74, 76f., 81f., 89, 96, 109, 215
— width 12f., 76, 94
Bayard-Alpert-Gauge 3
Be-window 70
beam-Bragg *129*–192, 203ff.
— molecular 213
bending-band 72
binding energy 29f., 37, 53, 55, 67ff., 71f., 76, 90ff., 96, 194, 212, 224, 227
bombardement-electron 8, 12f., 17, 21, 88, 137
— ion 6f.
bond-adsorption 79, 125, 194, 207, 212, 216, 230, 235
— chemical 30, 33, 77, 79, 81, 96, 185, 187, 207, 213, 220
— chemisorption 33, 77, 79, 123, 126, 235
— surface 64, 79
— vibrational 207
Bragg-beam *129*–192, 203ff.
— equation 140, 143
— peak *129*–192, 193, 198ff., *201*–202
— peak-shift 193, 201f.
Bravais lattice 130f.
bremsstrahlung 69, 86ff., 93
de Broglie-property 9, 129, 133

bulk-diffusion 4 ff.
— plasmon 58, 60, 62
βspectrometer 67

C-window 70
capacitance 10 f., 14, 20, 85, 88, *120*–121, 124
catalysis 1 f., 4 f., 23, 186, 207, 211, 232 f., 235 f.
CHA-cylindrical hemisphere analyzer *14*–15, 71
— resolution 15, 17, 23
channel-plate 71
channeltron 71
characteristic-losses 9, 22, 38, 48 f., 53 f., 59, 95,
   193
— radiation 24, 69 f., 86
charging effect 19, 72, 187
charge-distribution 58, 115, 123 f., 126, 156, 212
chemical bond 30, 33, 77, 79, 81, 96, 185, 187, 207,
   213, 220
— effect *33*–34, 57, 85, *96*–97
— shift 33 f., 62, 67 ff., 79, 91, 96 f.
— state 55, 57, 62, 91
chemisorption bond 33, 77, 79, 123, 126, 235
— state 31, 64, 67, 71, 76, 79, 113, 212, 215
Clausius-Clapeyron-equation 212
cleaving 5, 82, 106, 147, 168, 186 ff., 220
CMA-cylindrical mirror analyzer *12*–13, 15, 17,
   21 ff., 46 f., 54
— resolution 12 ff., 17, 23
coadsorption *230*–232
coefficient-backscattering 39
— reflection 175, 198
— secondary electron emission 23, 85
— sticking 2, 4, 221
— thermal expansion 201 f.
— transmission 77, 175, 198
coherence 147, 149, *159*–160, 162, 165, 170, 232
coherent-potential approximation 78, 94 f.
coherent-structure 223, 225 f.
coincidence lattice 132 f., *149*–152
collective excitations (→ phonons, → plasmons,
   → vibrations)
collector 9 ff., 20, 22, 54, 60, 63, 71, 86 ff., 99,
   111, 119, 121, 136, 139
competitive adsorption 138, 232, 235
concentric hemisphere analyzer (→ CHA)
condensation 6
conduction electrons 53, 57, 95
constant-dielectric 58, 63
— force 193, 197, 201
— lattice 182 f., 193
— optical 43, 60, 76, 207
contact potential *115*–127
convolution-integral *47*–49, 90, 93, 110, 112
cooperative adsorption 231
core-electron 9, 29, 53, 55, 68 f., 90 f., 96
— hole 17, 24 ff., 34, 85, 89 f., 92

— ion 173, 176
— level 30 f., 33, 70, 72, 85, 88 f., *91*–92, 94, 96
— state 24, 72, 85, 90
corrosion 1, 91
Coster-Kronig-transition 26, 32, 35, 92
Coulomb-interaction 27, 57
counting-ion 42
coupling-j-j 27 f., 30
— L-S 27 f.
— plasmon 85, *95*, 96
cross section-ionization 18, 34, 36, 38 f., 46, 57,
   214
— photodesorption 215
— photoionization 18
— reaction 213
cryopump 3
Curie-temperature 60, 78
current-photo 86 f.
cylindrical mirror analyzer (→ CMA)

d-band 76, 78 ff., 91 ff., 97, 102, 113 f., 230
Debye-spectrum 195
— temperature 195 f., 199 ff., 205
— Waller-factor 175, 178, 194 f., 198, 204 f.
decomposition 3, 57, 186, 227, 233
deconvolution (→ convolution) *47*–49
defects-structural 1, 130, 159, 170, 173, 184, 233
definition-surface 1
density of states 31, 49, 55, 67 f., 74 f., 77, 79, 82,
   89 ff., 93 ff., 97, 99, 110
— adsorbate 24, 33, 53, 55, 81
— joint optical d. o. s. 75
depth-escape 8, 37 ff., 44, 67
— penetration 8, 18, 44, 59 f., 85, 157 f., 175,
   196 ff., 207
— profiling 45 f.
desorption-electron stimulated 6, 8, 18, 21, 46,
   111, 213 ff.
— field 5, 85
— kinetics 194, 212
— laser induced 47
— probability 214
— thermal flash 42, 207, 211 f., 218, 224
dielectric constant 58, 63
— function 59
— theory 59 f.
differentiation-technique (→electronic modulation
   technique) 5, *9*–12, *17*–24, 54 f., *86*–88
— angle 138, 167, 180, 182
diffraction-elastic 9, 62, *129*–192, 195
— inelastic 9, 54, *60*–62, 64
— intensity pattern *129*–192
— pattern *129*–192, *133*–135
— pattern analysis *145*–146, 151 f., 156, 171,
   176, 178 f., 181
— spot *129*–192

diffraction-elastic, x-ray   9, 129, 135, 149, 152, 157, 159 f., 175, 181, 183, 194 f., 196
diffusion-bulk   4 ff.
— pump-mercury   3
— pump-oil   3
— surface   5, 47, 161, 164 f., 201, 212, 216 ff., 234
diode method   *121*–122
dipole-interaction   35, 125, 221, 226
— moment   115 f., 122 ff., 207, 221, 223 f.
— radiation   25
— scattering   193 f., 205 ff.
direct transition   75
discharge lamp-gas   3, 70
disordered structure   *159*–165, 166, 170
dispersion-analyzer   12 f., 15, 70
— plasmon   62
— x-ray   85
displacement-mean-square   193 ff., 199
display type-LEED   135, *137*–140
dissociation
distribution-angular   38, 54, 61, 79, 129
— charge   58, 115, 123 f., 126, 156, 212
— curve-energy   53 ff., *67*–84
— energy   8 ff., 12, 14, 17 f., 20, 22, 26, 29, 31, 47 ff., 54, 62, 64, 67, 71, 74, 90, 99, *100*–103, 111, 212
— external   76, 110
— internal   75 f., 110
— total energy (→ TED)
domain   146 ff., 151, 159, 161 ff., 168 ff., 184 f., 188, 218, 220, 225 f., 232 f., 235
dynamical theory   2, 62, 129 f., 140, 154, 157, 170, *171*–179, 180, 183, 197, 199, 205

EELS (→ spectroscopy-EELS-electron energy loss spectroscopy)
effect-charging   19, 72, 187
— chemical   *33*–34, 57, 85, *96*–97
— photo   24
— size   *160*–161
elastic-diffraction   9, 62, *129*–192, 195
— scattering   38, 48, 53, 60, *129*–192, 195 f., 202 f., 205
electrochemical potential   115 f.
electromagnetic radiation   8, 25, 67, 70
electron-Auger   5, 9, *17*–51, 53, 55, 68 f., 72, 85, 110
— bombardement   8, 12 f., 17, 21, 88, 137
— conduction   53, 57, 95
— core   9, 29, 53, 68 f., 90 f., 96
— diffraction-elastic   9, 62, *129*–192, 195
— diffraction-inelastic   9, 54, *60*–62, 64
electronegativity   122 f.
electron-emission of secondary   8 ff., 12, *17*–51, 53, 72, 76 f., 85, 87 f., 138, 186
— emission of thermionic   *117*–118

— energy loss spectroscopy (→ EELS)
— field emission   5, *8*, 18, *99*–107, *118*–119, 124
— gas-free   57 f., 76, 100, 102, 106, 123
electronic-factor   228
electron modulation-technique   5, *9*–12, *17*–24, 54 f., *86*–88
— multiplier   12, 14, 19 f., 22, 24, 69, 71, 99
— phonon-interaction   48, 175, *193*–209
— photo   8, 25, *67*–85, 106, 118
— quasielastic   9, 53 f., 175, 203 f., 212
— secondary (→ emission of secondary electrons)
— stimulated desorption   6, 8, 18, 21, 46, 111, 213 ff.
— valence   9, 30, 33, 38, 53 f., 57, 60, 62, 68 f., 74, 79, 104, 122, 126, 174 f., 178, 218, 220
electrostatic interaction   58, 69
Eley-Rideal mechanism   223 ff.
ellipsometry   42 f., 211, 220
emission-Auger electrons   5, 9, *17*–51, 53, 55, 68 f., 72, 85, 110
— field-electron   5, *8*, 18, *99*–107, *118*–119, 124, 212, 214
— photo-electron   8, 25, *67*–85, 106, 118
— secondary-electron   8, 12, *17*–51, 53, 72, 76 f., 85, 87 f., 138, 186
— thermionic electron   *117*–118
— x-rays   17, 25, 34, 41, 49, 85 ff., 92, 94
energy-adsorption (→heat adsorption)
— APS   *91*–92
— Auger   11, 20, *30*–34, 41, 48
— binding   29 f., 37, 53, 55, 67 ff., 71, 76, 90 ff., 96, 194, 212, 214, 227
— characteristic energy losses   9, 22, 38, 48 f., 53 f., 59, 95, 193
— distribution   8 ff., 12, 14, 17, 20, 22, 26, 29, 31, 47 ff., 54, 62, 64, 67, 71, 74, 90, 99, *100*–103, 111, 212
— distribution curve   5 ff., *67*–84, 111
— distribution-total (→TED)
— profile   61
— shift   60, 62 f., 96 f., 104 ff., 126, 158, 174, 198, 201
— surface-free   166, 193
— threshold   76 f., 85 ff., 96 f., 119, 215
enhancement factor   104 ff.
entropy   194, 216, 218
epitaxy   2, 6, 41, 133, 184 f., 211, 221, 225
ESCA (→ spectroscopy-ESCA (electron spectroscopy for chemical analysis))
ESCA-spectrometer   14 f., 69, 211
escape depth   8, 37 ff., 44, 67
evaporation   3, 6, 18 f., 39 f., 42 f., 129, 136, 138, 207, 211 f., 221, 230
Ewald construction   *143*–145, 157 f., 166 ff.
excitation rate   88, 90

expansion-thermal   193, 198, 201
external distribution   76, 110

facets   *166*–170, 173
factor-attenuation   38, 175
— Debye-Waller   175, 178, 194, 198, 204 f.
— electronic   228
— enhancement   104 ff.
— preexponential   194, 232
— scattering   152 f., 157, 170 f., 173 ff., 180 f., 185
— structural   228
— structure   153, *156*–157, 170 f., 181, 195
Faraday cup   54, 99, 129 f., 138 ff., 205
Fermi-Dirac statistics   100
Fermi level   9, 36, 53 f., 71 f., 74 f., 79, 81, 88 ff.,
   94 f., 97, 100, 102, 106, 112 f., 115, 119, 230
FERP (→ field emission retarding potential
   method)
field-desorption   5, 85
— electron emission   5, 8, 18, *99*–107, *118*–119,
   124, 212, 214
— electron energy spectrometer   *99*, 102, 105
— electron energy resolution   99, 102, 105
— emission energy distribution curve   81
— emission microscope   99, 119, 212, 214
— emission retarding potential method, FERP
   119
— emission spectroscopy (→ spectroscopy-field
   emission)
— ion microscope   99
— ion spectroscopy (→ spectroscopy-field ion)
films-evaporated   6, 19 f., 39, 42, 72, 136, 207,
   211 f., 230
— thin   6, 60, 62 f.
flash-desorption   42, 207, 211 f., 218, 224
fluorescence screen   10, 21, 87, 130, 136, 139 f.,
   145
— x-ray   35, 85
— yield   35 f., 80, 90, 92
force-constant   193, 197, 201
— van der Waals   220
Fowler equation   118
Fowler-Nordheim equation   100, 118
free electron gas   57 f., 76, 100, 102, 106, 123
Frenkel equation   216
frequency-threshold   118

gas-discharge lamp   3, 70
— free-electron   57, 76, 100, 102, 106, 123
— lattice   217 f., 226
— residual   1, 3, 7
gauge-Bayard-Alpert   3
— ionization   3, 4
geometric theory   *140*–152, 154, 156, 179
getter pump-ion   3, 7
glancing angle   17 ff., 22, 44, 87
Grüneisen relation   201

de Haas-van Alphen effect   75
H₂-resonance radiation   70
He-resonance radiation   67, 70, 76, 79
heat-adsorption   165, 211 f., 216, 218, 220 f.,
   226 ff., 230 f., 234
— specific   193, 195, 201
height-peak-APS   91, 93, 97, 211
— peak-Auger   23, 34, 40 ff., 234
hemispherical analyzer (→ analyzer-CHA)   *14*–
15, 71
— resolution   *14*–15
high resolution energy loss spectroscopy (→
   spectroscopy-hrels)
hole-core   17, 24 ff., 34, 85, 89 f., 92
127°-analyzer   *13*–14, 17, 19 (→ analyzer-127°)
— resolution   *13*–14, 20

ILEED   9, 54, *60*–62, 64
imperfections (→ defects)
incidence angle   18, 21, 40 f., 133, 188 f.
incoherent-structure   132 f., *149*–152, 165, 184,
   218, 223 f.
indirect transition   75
inelastic-diffraction   9, 54, *60*–62, 64
— scattering   53, 60, 64, 129, 138, 149, 193
inner potential   158, 174 f., 178, 180, 201
INS (→ spectroscopy-INS (ion neutralization
   spectroscopy))
insulator   19, 41, 72, *187*–188
intensity-Auger   22, *34*–39, 42, 220
— bulk-plasmon   60, 62 f.
— LEED   41 f., 62, *129*–192, 193, *194*–200, 201,
   218 f., 221 ff.
— measurements   130, 135, 158 ff.
— pattern   *129*–192
— photoemission   94
— profile   61 f., 134
— surface plasmon   58 ff., 62 ff.
— x-ray   85, 88, 90, 93
interaction-Coulomb   27, 57
— dipole   35, 125, 221, 226
— electron-phonon   48, 175, *193*–209
— electrostatic   58, 69
— intermolecular   163 f., 212, *216*–220, 226,
   230–232, 233 ff.
interband-transition   33, 53, 59 f., 95
intraband-transition   53, 59
interference-function   154 f., 159 f., 162, 169 ff.
intermolecular interaction   163 f., 212, *216*–220,
   226, *230*–232, 233 ff.
internal distribution   75 f., 110
ion-bombardment   6 f.
— core   173, 176
— counting   42
— getter pump   3, 7
— sputtering   6

ionicity 69
ionization-cross section 18, 34, 36, 38 f., 46, 57, 214
— gauge 3, 4
— losses 41, 55
— potential 67, 104
— secondary 18
— spectroscopy (→ spectroscopy-IS)
IR-radiation 207
— spectroscopy (→ spectroscopy-IR)
IS (→ spectroscopy-IS)
Ising model 218
isotherms-adsorption 221
I-V intensity-voltage curve *157*–159, 170 f., 173 ff., 176 ff., 183, 193, 198 f., 201

j-j-coupling 27 f., 30
joint optical density of states 75

Kelvin method (→ vibrating capacitor method)
kinematical theory 62, 140, 149, *152*–159, 163, 166, 170 f., 173, 175 f., 179 ff., 183, 194, 199, 204, 218
kinetics 21, 23, 41, *46*–47
kinetics-adsorption 83, 212, 214, 221
— desorption 194, 212

lamp-discharge 3, 70
Langmuir model 83
Langmuir-Hinshelwood mechanism 233
laser induced desorption 47
lattice Bravais 130 f.
— coincidence 132 f., *149*–152
— constant 182 f., 193
— gas 217 f., 226
— reciprocal *141*–143
Laue equation 143, 154, 157 f., 168, 171
LEED 9 f., 12 f., 17, 19 ff., 34, 39, 41 f., 53 f., 60, 62, 87, 111, 114, 121 f., *129*–192, 193 f., 196, 211 ff., 216, 218, 220, 222, 225 ff., 231 ff.
— display-type 135, *137*–140
— intensity 41 f., 62, *129*–192, 193, *194*–*200*, 201, 218 f., 221 ff.
Lennard-Jones 6–12 potential 193
level-core 30 f., 33, 70, 72, 85, 88 f., *91*–92, 94, 96
— Fermi 9, 31, 53 ff., 71 f., 74 f., 79, 81, 88 ff., 94 f., 97, 100, 102, 106, 112 f., 115, 119, 230
LiF-window 67
lifetime 26, 32 f., 76, 89, 92 ff., 215
lock-in-amplifier 10 ff., 21 f., 71, 86, 88, 121
losses-characteristic 9, 22, 38, 48 f., 53 f., 59, 95, 193
— ionization 41, 55
— mechanism 33
— peak-shape 53, 57

— peak-width 53 ff.
— phonon 48, 175
— plasmon 22, 33, 41, 45, 48, 53, 55, *57*–60, 62 f., 72
— probability 193
— profile 61 f.
L-S-coupling 27 f.

manipulator 19, 135, *136*–137
mass spectrometer 4, 212 ff., 233, 235
— omegatron 4
— quadrupole 4
mean-free path 2, 7 f., 39 f., 44, 76, 79, 175
— square displacement 193 ff., 199
mechanism-APS *88*–91
— Auger *24*–30
— Eley-Rideal 223 ff.
— Langmuir-Hinshelwood 233
— loss 33
melting 184 f.
mercury-diffusion-pump 3
metallurgy 23, 41
metals-transition 33, 67, 69, 97, 102, 104, 222, 225
method-averaging 152, 154, *179*–182
— diode 121–122
— one electron 33, 95
microanalysis 18
microbalance 42 f., 211
microscope field emission 99, 119, 212, 214
— field ion 99
— scanning electron 24
minimum polarity model 94
model-Langmuir 83
— Ising 218 f.
— minimum polarity 94
— rigid band 78, 94
— three step 74, 77, 79
— two step 62
modulation technique 5, *9*–12, *17*–24, 54 f., *86*–88
molecular beam 213
— surface orbital 64
— turbo pump 3
monochromator-x-ray 70, 77
Mößbauer-spectroscopy (→ spectroscopy-Mößbauer)
muffin-tin-potential 174
multichannel analyzer 23, 71, 211
multiple scattering 141, 149 ff., 154, 159, 173, *176*–179, 180 ff.
multiplier electron 12, 14, 19 f., 22, 24, 69, 71, 99

Na-source 70
Ne-resonance radiation 70
noble gas ion sputtering 6 f.
noise-shot 12 f., 87

off axis filament   136
oil diffusion pump   3
oil rotary pump   3
omegatron mass spectrometer   4
one electron method   33, 95
— process   56, 69, 95, 99
optical path   76
order-disorder transition   170, 212, 225
ordered adsorbed phases   *216*–220, 129–192
oxidation   5, 33f., 45f., 69, 80, 85, 96f., 129, 211, 234, 236
oxide   58, 72

patches   115, 118
path-mean free   2, 7f., 39f., 44, 76, 79, 175
— optical   76
pattern-diffraction   129–192
peak-Bragg   129–192, 193, 198ff., 201
— Bragg shift   193, 201f.
— shift energy   62, 96, 104f., 126, 158, 174, 198, 201f.
penetration depth   8, 18, 44, 59f., 85, 157f., 175, 196ff., 207
periodic structure   129–192
PES (→ spectroscopy-photoelectron)
phonon (→ vibrations)   9, 48, 53, 175, 193–203
— electron interaction   48, 175, *193*–209
— losses   48, 175
— scattering   48, 53, 203f.
— surface   193f.
photo-current   86f.
— desorption cross section   215
— effect   24
— electron   8, 25, *67*–85
photo-electron emission   8, 25, 67–85, 106–118
— electron spectrometer   14, 69–72, 85
— emission intensity   94
— ionization cross section   18
— yield   87
photometer, spot   54, 139
photon   18, 24, *67*–85
physical adsorption   220f.
plasma oscillation (→ plasmon)
plasmon-bulk   58, 60, 62
— coupling   85, 95f.
— dispersion   62
— intensity   59, 60, 62f.
— losses   22f., 41, 45, 48, 53, 57–60, 62
— surface   58f., 62f.
— transition   33
— volumen (→ bulk)
potential-contact   115–127
— electrochemical   115f
— inner   158, 174f., 178, 180, 201
— Lennard-Jones   193
— muffin-tin   174

— scattering   174ff.
— surface   116
preexponential factor   194, 232
preparation   4–7, 147, 186
probability-desorption   214
— loss   193
— scattering   194
— transition   31, 75, 90, 110
— tunneling   100f., 104, 118
process-one electron   59, 69, 95, 99
— two electron   69, 95, 99, 109f.
profile-angular   61
— loss   61f.
profiling-depth   45f.
proton model   81
pump, cryo   3
— ion getter   3, 7
— mercury diffusion   3
— oil diffusion   3
— rotary oil   3
— sorption   3
— sublimation   3
— turbo molecular   3

quadrupole mass spectrometer   4
qualitative analysis   41
quantitative analysis   19f., 39, 41–47
quasielastic electrons   9, 53f., 175, 203f., 212

radiation, characteristic   24, 69f., 86
— dipole   25
— electromagnetic   8, 25, 67, 70
— H_2 resonance   70
— He resonance   67, 70, 76, 79
— infrared   207
— Ne resonance   70
— synchrotron   67, 70, 82
— ultraviolet   67–84
— x-ray   17f., 24, 69, 85, 92, 94, 157
rate-excitation   88, 90
reaction 232–236
— cross section   213
reconstruction   114, 173, 185, 187
reciprocal lattice   141–143
reduction   5
reflectance spectroscopy   207
reflection-coefficient   175, 198
relation, Grüneisen   201
residual gas   1, 3, 7
resolution-CHA   15, 17, 23
— CMA   12ff., 17, 23
— field electron energy spectrometer   99, 102, 105
— RFA   9ff., 17, 20
— x-ray spectrometer   70f., 77
— 127°   13f., 20

resonance tunneling 99f., *103*–106, 109, 118, 124
retarding field analyzer (RFA) 9–12, 17, 20ff., 46, 49, 54, 67, 71, 99
— resolution 9ff., 17, 20
Richardson-Dushman equation 147
rigid band model 78, 94
rotary oil pump 3
rules, selection 25

s-band 78, 93f., 113
scanning electron microscope 24
scattering-angle 180, 195
— back 8, 18, 38ff., 53f., 60ff., 64, 72, 129, 133, 138, 140, 172f., 176, 178, 196f., 199
— dipole 193f., 205ff.
— dynamical (→ dynamical theory)
— elastic 38, 48, 53, 60, *129*–192, 195f., 202f., 205
— factor 152f., 157, 170f., 173ff., 180f., 185
— inelastic 53, 60, 64, 129, 138, 149, 193
— multiple 141, 149ff., 154, 159, 173, *176*–179, 180ff.
— phonon 48, 53, 203f.
— single 149, 151f., 154, 173 (→ kinematical theory)
— thermal diffuse 139, 194, *202*–205
screen, fluorescence 10, 21, 87, 130, 136, 139f., 145, 172
secondary-electron emission 8ff., 12, *17*–51, 53, 72, 76, 85, 87f., 138, 186
— ionization 18
— yield 9, 19
selection rules 25
semiconductor 1, 5, 19, 22, 41, 46, 67, 72, 79, 82f., 131, 147, 156, 168, 185f.
shape, APS peak 91ff., 97
— Auger peak 11, 20, *30*–34, 41, 48
— characteristic x-ray's peak 94
— loss peak 53, 57
shift, Bragg peak 193, 201f.
— chemical 33f., 62, 67ff., 79, 91, 96f.
— energy 60, 62f., 96f., 104ff., 126, 158, 174, 198, 201
shot noise 12f., 87
signal-averager 23, 71
simple structure 132f., 150f.
single scattering 149, 151f., 154, 173 (→ kinematical theory)
size effect 160f.
soft x-ray spectroscopy (SXS) 85–98, 27, 67
soft x-ray appearance potential spectroscopy (SXAPS) 85–98
sorption pump 3
specific heat 193, 195, 201
spectrometer-AEAPS 87
— APS 86–88

— Auger 18–24, 31, 33, 68f.
— ESCA 14f., 69, 211
— mass 4, 212ff., 233, 235
— PES 14, 69–72, 85
— UPS 3, 67–84
— β 67
— 127° 13f., 17, 19
spectroscopy-AEAPS 85–98
— APS 8, 85–98
— Auger 5, 9, *17*–51, 54, 111, 137f.
— EELS 9, *53*–66, 212, 215
— ESCA 67–84
— field emission 99–107
— field ion 211
— high resolution energy loss 194, 205–207
— INS 8, 79, 109–114, 212
— IR 194, *207f.*, 211
— IS 53–57
— isochromat 85
— molecular 67, 70
— Mößbauer 194
— PES 8, 14, 24, 26, 33, 64, *67*–84, 95, 102, 109, 111, 211f.
— reflectance 207
— SXAPS 85–98
— SXS 27, 67, 85–98
— UPS 3, 67–84, 109f.
— XPS 27, 67–84, 109
spectrum, Debye 195
spherical deflector analyzer (→ CHA)
splitting, spot 146, 162f., 165, 168ff., 171, 184
spot-photometer 54, 139
— diffraction 129–192
— splitting 146, 162f., 165, 168ff., 171, 184
sputtering 5ff., 38, 44ff., 48
state-adsorption 81, 114, 207, 214f.
— chemical 55, 57, 62, 91
— chemisorption 31, 64, 67, 71, 76, 79, 113, 212, 215
— core 24, 72, 85, 90
— density of 31, 49, 55, 67f., 74f., 77, 79, 82, 89ff., 97, 99, 110
— density of adsorbate 24, 33, 53, 55, 81
— joint optical density of 75
— surface 64, 79, 81ff., 102, 106, 115, 186
— vibrational 71, 174, *193*–202, 205, 212
— virtual bound 78f., 230
steps *166*–170, 183
sticking coefficient 2, 4, 221
structural analysis 9, 152, 181f.
— defects 1, 130, 159, 170, 173, 184, 233
structure, band 29, 31, 47–49, 67–84, 93ff., 102, 174, 176f.
— coherent 223, 225f.
— disordered 159–165, 166, 170
— factor 153, 156f., 170, 181, 195

structure, band, incoherent   132f., 149–152, 165, 184, 218, 223f.
— perodic   129–192
— simple   123f., 150f.
— super   184, 186
— surface   129–192, 223f.
sublimation pump   3
superstructure   184, 186
surface, band   82
— bond   64, 79
— definition   1
— diffusion   5, 47, 161, 164f., 201, 212, 216ff., 234
— energy   166, 193
— entropy   194, 218
— phonons   193f.
— plasmon   58ff.
— potential   116
— state   64, 79, 81ff., 102, 106, 115, 186
— structure   129–192, 223f.
— vibrations   9, 53, 182, *193–209*
SXAPS (→ spectroscopy-APS (soft x-ray appearance potential spectroscopy))
SXAPS-yield   88
synchrotron radiation   67, 70, 82
technique-convolution   *47–49*, 90, 93, 110, 112
— differentiation   5, *9–12*, *17–24*, 54f., *86–88*
— electron modulation   5, *9–12*, *17–24*, 54f., *86–88*
— molecular beam   213
— UHV   1, *2–4*, 85

TED   *100*–103, 104ff.
temperature-Curie   60, 78
— Debye   195f., 199ff., 205
— dielectric   59f.
theory-dynamical   2, 61, 129f., 140, 154, 157, 170, *171–179*, 180, 183, 197, 199, 205
— geometrical   *140*–152, 154, 156, 179
— kinematical   62, 140, 149, *152–159*, 163, 166, 170f., 173, 175f., 179ff., 183, 194, 199, 204, 218
thermal expansion   193, 198, 201
— coefficient   201f.
— diffuse scattering   139, 194, *202–205*
— flash desorption   42, 207, 211f., 218, 224
— vibration   159, *193–209*
thermionic electron emission   *117–118*
thin films   6, 60, 62f.
three step model   74, 77, 79
threshold energy   76f., 85ff., 96f., 119, 215
— frequency   118
titanium sublimation pump   3
total energy distribution (→TED)
tracer method   42, 211, 220
transition-Coster-Kronig   26, 32, 35, 92
— direct   75

— indirect   75
— interband   33, 53, 59f., 95
— intraband   53, 59
— metal   33, 67, 69, 91ff., 97, 102, 104, 222, 225
— order-disorder   170, 212, 225
— plasmon   33
— probability   31, 75, 90, 110
— transmission-coefficient   77, 175, 198
— function   38, 76
two step model   62
two electron process   69, 95, 99, 109f.
tubes-x-ray   18, 67, 69f., 79
tunneling   99f., *103*–106, 109, 118, 124, 215
— probability   100f., 104, 118
turbo molecular pump   3

UHV-technique   1, *2–4*, 85
uncertainty principle   26, 32, 92
UPS (→ spectroscopy-UPS (Ultraviolet photoelectron spectroscopy))
UPS-spectrometer   3, 19, *69–72*, 85
— yield   70
UV-radiation   *67–84*

valence   1, 29
valence-band   29ff., 49, 53, 72, 74, 76f., 81f., 89, 96, 109, 215
— electron   9, 30, 33, 38, 53f., 57, 60, 62, 68f., 74, 79, 104, 122, 126, 174f., 178, 218, 220
vibration-bond   207
— states   71, 174, *193–202*, 205, 212
— surface   9, 53, 182, *193–209*
— thermal   159, *193–209*
virtual bound state   78f., 230
vibrating capacitor method   *120–121*

van der Waals forces   220
window-Be   70
— C   70
— LiF   67
width-peak-APS   92, 94, 97
— Auger   26
— loss   53ff.
Wood-nomenclature   131, 146
work-function   9, 29ff., 71f., 74ff., 79ff., 88, 91, 99, 104, 110, *115–127*, 208, 211f., 214, 221ff., 225, 228f., 230f.
— measurement   42, 80f., 99, 115, *117–121*

x-ray-analysis   134, 140f.
— diffraction   9, 129, 135, 149, 152, 157, 159f., 175, 181, 183, 194f., 196
— dispersion   85
— emission   17, 25, 34, 41, 49, 85ff., 92, 94

x-ray-analysis,
— fluorescence   35, 85
— intensity   85, 88, 90, 93
— monochromator   70, 77
— radiation (→ AES-, APS-, PES-spectroscopy)
  17ff., 24, 69, 85, 92, 157
XPS-tubes   18, 67, 69f., 79
— yield   35f., 88, 90, 92

yield-Auger   18, 29, *34*–36, 85
— fluorescent   35f., 80, 90, 92
— photo   87
— secondary   9, 19
— SXAPS   88
— UPS   70
— XPS   35f., 88, 90, 92
Yttrium-source   70

Index of Substances and Adsorption Systems

Ag   4, 7, 39f., 42ff., 78, 123, 177f., 183, 187, 198, 201, 204f.
Ag–CO   228, 230
Ag–hydrocarbons   220
Ag–J   178f.
Ag–noble gases   220
Ag–Pd   43ff., 78, 185, 229
Ag–Pd–CO   229
Ag–W   42f.
Al   28, 33, 59f., 62ff., 69f., 73, 77, 123f., 178, 183, 197, 199f.
Al$_2$O$_3$ 33
Al–O   33, 63f.
Al–Si   168
Ar   6, 45, 137
Ar–Nb   220
Au   3, 7, 39f., 77, 121, 123, 184, 197, 233
Au–Cl   184
Au–CO   208
Au–Cu   185

B   95
Ba   123
Ba–Mo   225
BaTiO$_3$ 187
Ba–O   81
Ba–W   105f.
Be   7, 20, 33, 59, 70, 123
Be–O   33
Bi   184, 187
Bi–stainless steel   45
Br   28

C   4f., 7, 21, 33f., 38f., 45, 56, 64, 70, 91, 95f., 161, 187, 213, 220, 225
C–Mo   34
C–Ni   34, 227
C–Pt   233
C–W   55f.
C–Xe   220
Ca–W   105f.
Cd–Mo   49
Cd–TiO$_2$ 49
Cl–Au   184
Cl–Si   46, 186
Co   92, 94, 123
CO   4f., 7, 67f., 79, 213, 225, 231
CO–Ag   228, 230
CO–Ag/Pd   229
CO–Au   208
CO–Cu   126, 160, 208, 228, 230, 233
CO–Cu–O   233
CO–Mo   34

CO–Ni   64f., 67, 79f., 114, 207, 227ff.
CO–Pd   122, 139, 148, 225ff., 230f., 234f.
CO–Pd–H   231
CO–Pd–O   232, 234ff.
CO–Pt   235
CO–Pt–O   235
CO–Ti   81
CO–W   81f., 106, 207, 214f.
CO$_2$ 55
CO$_2$–Pt   235f.
CO$_2$–Pt–S   236
CO$_2$–Ti
Cr   45, 88, 91f., 94, 96f., 123, 198, 200ff.
Cr–O   96f., 168
Cs   39, 83, 123f.
Cs–GaAs   83
Cs–Ge   39
Cs–Ni   224f.
Cs–Si   39f.
Cs–W   225
Cu   3, 38f., 45f., 67, 74, 78f., 91, 94, 113, 119, 123f., 169f., 183, 197ff., 201
Cu–Au   185
Cu–CO   126, 160, 208, 228, 230, 233
Cu–CO–O   233
Cu–H$_2$S   46
Cu–Ni   43f., 78, 94f., 185, 229
Cu–O   147f., 233f.
Cu–O–S   234
Cu–Pb   185
Cu–S   234
Cu–Xe   221

Fe   7, 18, 46, 91f., 94, 123
Fe–Si   18, 40

GaAs   82, 168, 186
GaAs–Cs   83
Ge   35, 43, 82, 113, 169f., 185f., 220
Ge–Cs   39
GeH$_4$ 232
Ge–H$_2$O   186
Ge–K   39
Ge–O   232

H   4f., 35, 41, 70, 129, 157, 182, 211, 225
H–Ni   64, 81, 173, 214, 233
H–O–Ni   233
H–Pd   81, 231
H–Pd–CO   231
H–Ti   81
H–W   82, 173, 218f., 233
H$_2$O   220

H₂O–Ge 186
H₂S–Cu 46
He 3, 40, 67, 70, 76, 79, 109, 112
n-heptane–Pt 233
Hg 30
hydrocarbons–Ag 220

In 47f.
Ir 184

J–Ag 178f.

K–Ge 39
K–Ni 224
K–Si 39
Kr 26

Li 28, 123f.

Mg 58, 69f., 173
mica 6, 187f.
Mn 33f., 92, 123
MnO 33f.
Mn–O 33f.
Mo 5, 7, 38ff., 198, 201
Mo–Ba 225
Mo–C 34
Mo–Cd 49
Mo–CO 34
Mo–O 49

N 38f., 88, 136
N–Ni 64
N–Ti 81
N–W 82, 233
N–W–H 231
Na 28, 38f., 70, 124
Na–W 225
Nb 169
Nb–Ar 220
Ne 70
NH₃ 231
Ni 5, 7, 34, 41, 45, 60, 64, 78ff., 91f., 94, 112f.,
    123, 129, 161, 178, 180, 183, 196ff., 201, 203ff.
Ni–C 34, 227
Ni₃C 34, 232
Ni–CO 64f., 67, 79f., 114, 207, 227ff.
Ni–Cs 224f.
Ni–Cu 43f., 78, 94f., 185, 229
Ni–H 64, 81, 173, 214, 233
Ni–H–O 233
Ni–K 224
Ni–N 64
Ni–NiO 166
Ni–O 64, 67, 79ff., 113f., 173, 179, 227, 233
NiO 80f., 187

Ni–O–S 233
Ni–S 42, 64, 113f., 179, 233
Ni–Se 113f., 179
Ni–SO₂ 234
Ni–Te 179
Ni–Ti 94
noble gases 3, 6f., 45, 109
noble gas–Ag 220

O 4f., 21, 23, 33, 38f., 46f., 55f., 88, 91, 95, 97,
    113f., 182, 225
O–Al 33, 63f.
O–Ba 81
O–Be 33
O–CO–Cu 233
O–Cr 96, 168
O–Cu 147f., 233f.
O–Cu–S 234
O–Ge 232
O–H–Ni 233
O–Mn 33f.
O–Mo 49
O–Ni 64, 67, 79ff., 113f., 173, 179, 227, 233
O–Ni–S 233
O–Pd 148, 163ff., 235
O–Pd–CO 232, 234ff.
O–Pt 184, 234ff.
O–Pt–CO 235
O–Pt–S 234, 236
O–Rh 168, 181
O–Si 206f.
O–Sr 81
O–Ta 33, 85
O–Ti 62f.
O–TiO₂ 49
O–V 33f.
O–W 55f., 168, 173, 215
Os 123

P 10, 45f.
Pb 184, 197
Pb–Cu 185
Pd 4f., 42ff., 78, 91, 94, 123, 160, 197
Pd–Ag 43ff., 78, 185, 229
Pd–Ag–CO 229
Pd–CO 122, 139, 148, 225ff., 230f., 234f.
Pd–CO–H 231
Pd–CO–O 232, 234ff.
Pd–CO₂ 234, 235
Pd–H 81, 231
Pd–O 148, 163ff., 235
Pd–S 236
Pd–Xe 42, 125, 151, 220f.
Pt 23, 123, 129, 169, 184, 197
Pt–C 233
Pt–CO 235

Pt–CO–O 235
Pt–n-heptane 233
Pt–O 184, 234f.
Pt–O–CO 235
Pt–S 234
Pt–S–CO 231
Pt–S–O 234, 236
Pt–SO$_2$ 234

rare earths 91
Re 169
Rh–O 168, 181

S 5, 23, 41f., 47, 91, 113ff.
S–Cu 234
S–Ni 42, 64, 113f., 179, 233
S–Pd 236
S–Pt 234
S–Pt–CO 231, 236
S–Pt–O 234, 236
Sb–stainless steel 45, 187
Sc 92
Se 113f., 179
Si 18, 29, 39ff., 43, 45f., 49, 82, 113, 159, 186, 206, 220
Si–Al 168
Si–Cl 46, 186
Si–Cs 39f.
Si–Fe 18, 40
Si–K 39
Si–O 206f.
SiO$_2$ 45f.
Sr 104
Sr–O 81
Sr–W 105
stainless steel 3, 72f., 86f., 207
— Bi 45
— Sb 45, 187

Ta 33, 85
Ta$_2$O$_5$ 121
Ta–O 33, 85
Te 186
Te–Ni 179

Th 91, 123
Ti 3, 62f., 92, 123
Ti–CO 81
Ti–CO$_2$ 81
Ti–H 81
Ti–N 81
Ti–Ni 94
Ti–O 62f.
TiO$_2$–O 49
TiO$_2$–Cd 49
toluene 233

U 123
UO$_2$ 168
V 33f., 92
V–O 33f.
V$_2$O$_5$ 187

W 5, 7, 38ff., 42f., 56f., 81f., 85, 102ff., 123, 176, 183, 205, 214
W–Ag 42f.,
W–Ba 105f.
W–C 55ff.,
W–Ca 105f.
W–CO 81f., 106, 207, 214f.
W–Cs 225
W–H 82, 173, 218f., 233
W–N 82, 233
W–N–H 231
W–Na 225
W–O 55f., 121, 168, 173, 215
W–Sr 105

Xe 42, 125, 171, 220
Xe–C 220
Xe–Cu 221
Xe–Pd 42, 125, 151, 220f.

Y 28
YMnO$_3$ 187

ZnSe 82
zeolite 39
Zr 28, 104